LIFE SCIENCES

Evolution and Growth

LIFE SCIENCES

Evolution and Growth

Archana Shukla

MD PUBLICATIONS PVT LTD
NEW DELHI
www.mdppl.com

Published by :

MD Publications Pvt Ltd
"MD House", 11, Darya Ganj,
New Delhi-110 002
Phone : +91-11-41563325 (Customer Service)
+91-11-41562846 (Editorial)
E-mail (orders) : order@mdppl.com
Website : www.mdppl.com

ISBN : 978-81-7533-191-4

Published and Printed by Mr. Pranav Gupta on behalf of **MD Publications Pvt Ltd** at Printext, New Delhi.

PREFACE

The Life Science also does much more. It not only shows the general interested public what a few great scientists were really like, how they worked and what they accomplished, it shows the scientists themselves how much they can gain through the self knowledge that they can discover in the history of science in the history of what they themselves have been doing. The author spent her life breaking down the barriers between the different learned disciplines and in spreading the information that has so often been isolated in cells where its importance is not realized. If science contains within itself the potentiality of danger and we have to admit that it does if, for example, the scientists do not quite know what to do with the genic they have let out of the atomic flask, and if the ordinary voter needs to know what the problems are that he sometimes settles on election day, then it becomes very urgent that we discover some way to direct our present store of knowledge to those who lack it, but whose decisions, either wise or foolish, will affect our future.

This is not to argue, however, that our vast store of technical information should be broadcast, or even that broadcasting it is possible, because here the scientists themselves are becoming swamped. There is another type of scientific knowledge, however, which is much more relevant to our social and philosophical problems, the kind of knowledge that the author herself knew so well. When science is viewed as a whole, its many details tend to fall into place and what we may call the scientific spirit begins to emerge. Those ethical and moral principles whose application to scientific activity have proved so profitable in the past become easy to recognize. One does not have to be a scientist to realize that an extension of these principles to other fields of human behavior would be universally beneficial.

Unfortunately the learned world of the twentieth century has been splitting right down the middle, and the chasm between the two halves seems to be widening. Unless we can mend this growing rift we are threatened with a severe case of schizophrenia indeed, Sir Charles Snow has recently called attention to what he has named the "two cultures" and these, of course, are the culture of the sciences and the culture we call the humanities. Scientists and humanists continue as always to make social and personal contacts but, on the level of their major interests, they tend to become strangers. Fortunately the seriousness of this division of our modern culture is now recognized, and efforts are being made to heal the fracture. The first rather tentative steps have already been taken in the field of the history of science, because in this field the two cultures have always co-existed. Here they have preserved their ancient unity and here both the scientist and the humanist can glimpse the great contributions to our civilization that both have made.

Our leading engineering institutes have become aware of the danger of graduating highly trained technical men who are nothing but highly trained technical men, and, as a first step in expanding their educational programs, they have established courses in the history of science that all students are required to take. Old established universities are creating new departments for teaching and for research in the history of science, and the demand for historians of science has recently become so great that there is now a marked shortage of trained personnel.

Those who discovered "truth", were recognized human beings, interesting in themselves and often intensely even cantankerously human. Science was a growing, living body of experimentally verifiable knowledge, fascinating because the universe itself is fascinating, and important because, at least by human standards, our species, which created science, is important. Thus, throughout all his published work, no matter how abstruse or technical, there was always a human and a living element.

The discovery of the role of DNA and its drawing together of all the life sciences has at long last made it possible for the layman to approach the subject of life, and to aspire to an understanding of the amazing knowledge science is gaining of it. It was this that emboldened me to undertake this book.

I wish to express my most sincere thanks to the scientists who were so kind as to open their material to me, and to read and criticize the various parts of this book. At a considerable cost in time to themselves and, in some cases, to their staffs, they have given me invaluable help.

I am very grateful to Mr. Pranav Gupta, Director of MD Publications Pvt Ltd, New Delhi for his cooperation and support for this book.

Archana Shukla

CONTENTS

1

Theories of the Generation of Life

Introduction

We may find a belief in the possibility of the spontaneous generation of living things amongst all peoples and at all times; beginning in remote antiquity and finishing in our own days. Even now, in the period of the blossoming of exact science in the culturally advanced nations, it is common for their ordinary inhabitants to be convinced that maggots arise from dung and rotting meat and that various domestic pests arise of their own accord out of rubbish, mud and dirt. These superficial observations miss the fact that dung and filth are to be found in those places where pests lay their eggs from which the new generation of living things develops.

Tremendous significance was attached to these everyday, uncritical observations of creation characteristic of ancient peoples, at a time when nature was still not studied in detail, nor submitted to analysis and dissection but was accepted in its entirety as the immediate perception of the intuition. In his book *Urzeugung und Lebenskraft*, e.o. v. Lippmann gives a wide range of material to show how extensively such beliefs were held. For example, in China in remote times people believed that aphids would grow by spontaneous generation on bamboos if the young shoots were planted out in warm moist weather. In the Indian holy books there are also references to the sudden appearance of various parasites, flies and beetles from sweat and dung. In the cuneiform writings of Babylon one may read that the mud of canals forms worms and other animals from its substance.

In ancient Egypt the view prevailed that the layer of silt left behind after the flooding of the Nile could give rise to living creatures when it was warmed a little by the sun. Frogs, toads, snakes and mice could originate in this way. In this case one might easily convince oneself by direct observation that the front part seemed already finished and alive while the hind part still consisted of undifferentiated damp earth.

We also find a repetition of these tales among the ancient Greeks (e.g. Diogenes Apolloniates) and in the writings of the famous Roman sage, Pliny. Such stories were widely current both in the East and the West, in the Middle Ages and far more recently. Shakespeare's audiences were not surprised when Lepidus, in *Antony and Cleopatra*, asserted that in Egypt crocodiles are produced from the mud of the Nile under the influence of the warm southern sun.

In general, it appears to be highly characteristic of the history of spontaneous generation that among diverse peoples living at different times and at different cultural stages, we almost always find stories of the spontaneous development of organisms of one kind or another. Here maggots arise from dung and rotting meat, here lice form themselves from human sweat, here fireflies are born from the sparks of a funeral pyre, and finally, frogs and mice originate from dew and damp earth. Wherever man has met with the unexpected and exuberant appearance of living things he has regarded it as an instance of the spontaneous generation of life. Among the ancient peoples the belief in spontaneous generation did not arise as a consequence of any particular philosophy. For them spontaneous generation was simply an obvious, empirically established fact the theoretical basis of which was of secondary importance.

The ancient teachings of India, Babylon and Egypt bound up the origin of life with various religious legends and traditions. From this point of view spontaneous generation was merely a particular manifestation of the creative will of gods or demons. But at the very source of our European culture in ancient Greece, on the replacement of theogony, a mystical interpretation of nature, cosmogony arises as the beginning of scientific investigation.

Although all the Greek philosophers from the Miletians to Epicurus and the Stoics acknowledged spontaneous generation as an incontrovertible fact, their philosophical treatment of this fact went far beyond the framework of the previous mystical presentations. They contained the beginnings of all the concepts which were developed later in connection with the question of the origin of life.

Even the earliest Greek philosopher, Thales, who lived from about 624 to 547 B.C., approached the problem of the essential nature and origin of life from an elementary materialist position. Thales and the other philosophers of the Miletian school recognised, as a fundamental principle, the objective existence of matter as something which is always living and always changing from the beginning of time. Life is inherent in matter as such. Thus, although the Miletians believed in the spontaneous generation of living things from mud, slime and such materials, they treated this phenomenon as the selfcreation of individual organisms, and not as one requiring the intervention of any special mystical force. This point of view was developed later by Empedocles (c. 485-425 B.C.), who held that plants and animals are formed from substances which, although not organised, are already living, either by birth from things like themselves or from things unlike themselves, i.e., by spontaneous generation. A particularly clear enunciation of the idea of the self-creation of living things is to be found in the works of Democritus (460-370 B.C.). In this doctrine ancient Greek materialism reached the height of its development although it had also already acquired a somewhat mechanistic character. According to the view of Democritus matter forms the basis of the universe and consists of a multitude of very small particles (atoms) which are in constant motion and are separated from one another by empty spaces. This mechanical motion of the atoms is inherent in matter, and on it depends the process of organisation of all individual objects. In particular, life appears, not from an act of divine creation, but as the result of the mechanical forces of nature itself. According to Democritus the primary development of living creatures, or their spontaneous development from water and mud, occurs when minute particles of moist earth come together with atoms of fire in a fortuitous but completely determinate way in the course of their mechanical movement. Another illustrious ancient Greek thinker, Epicurus (342-

271 B.C.) took up the same philosophical position a hundred years later. We may find an exposition of his views in the well-known poem of Lucretius Carus, De rerum natura. According to this source, Epicurus taught that, thanks to the moist heat of the sun and the rain, there arise from earth or manure, worms and a multitude of other creatures. But this happens without the participation of any spiritual influence whatever. Spirits, in the form of non-material forces, do not exist, according to Epicurus. The spirit is material and consists of small, very delicate and smooth atoms. The mechanical juxtaposition of atoms in empty space also leads to the formation of multifarious things, in particular, living beings. According to him, the cause of the motion of the atoms resides in matter itself and does not depend on any 'initial impulse' or other meddling of gods in the affairs of the world.

Thus, even hundreds of years before the beginning of our era, the phenomenon of spontaneous generation was explained materialistically by many schools of philosophers as being the self-creation of living things without the participation of any spiritual forces. The matter may be summed up historically by saying that the later development of the idea of spontaneous generation was bound up, not with the materialistic 'line' of Democritus but with the opposing idealistic 'line' of Plato.

Plato himself (427-347 B.C.) hardly concerned himself directly with the problem of spontaneous generation. In the Phaedo he only touches superficially on the question of the possibility of the formation of living things under the influence of warmth and decay. However, in complete harmony with his general philosophical position, he maintained that life is not inherent in plant and animal matter but this can only be brought to life by the infusion into it of the immortal spirit or Psyche.

This idea of Plato's played a tremendous part in the later development of the problem in which we are interested. It was reflected to some extent in the teaching of Aristotle which later formed the basis of the mediaeval scientific culture and dominated people's minds for nearly 2000 years.

Aristotle (384-322 B.C.) gave to mankind by far the broadest synthesis of the achievements of ancient science, embracing all the factual material which had been accumulated up till that time. He unfolded his views on the origin of life in a number of biological works concerning the origin of animals: Historia animalium, De partibus animalium, and De generatione animalium. According to Aristotle animals are born from others like themselves but equally, they arise and always have arisen by spontaneous generation from nonliving matter. He wrote as follows:

> *Such are the facts, everything comes into being, not only from the mating of animals but from the decay of earth and dung. And among plants the matter proceeds in the same way, some develop from seed, others, as it were, by spontaneous generation by natural forces; they arise from decaying earth or from certain parts of plants.*

Ordinary worms, the grubs of bees and wasps and also ticks, greenflies and various other sorts of insects arise, according to Aristotle, from dews in the presence of decaying mud and dung, from dry trees, hair, sweat and meat. All sorts of intestinal worms are formed from decomposing parts of the body and excreta. Midges, flies, moths, mayflies, dung beetles, cantharides, fleas, bugs and lice (partly as such and partly as grubs) arise from the slime of wells, rivers and seas, from the soil of the fields, from mould and dung, from rotting wood and fruit, the dirt of animals, from all sorts of filth, from the sediment of vinegar and also from old wool. Not only insects and worms but other living things can, according to Aristotle, arise by spontaneous generation. Thus crayfish and various molluscs originate from wet earth and decaying slime, eels and some other fishes from marine silt, sand and decaying water weeds. Even frogs, and under certain circumstances salamanders too, can arise from the curdling of slime. Mice arise from damp earth. Some higher animals also arise in a similar way, first manifesting themselves in the form of worms. "For this reason, and concerning human beings and quadrupeds", Aristotle wrote, "if they were sometimes earth-born, as some people maintain, one may postulate two methods of arising, either from worms which form themselves first, or from eggs."

However Aristotle did not merely describe various cases of spontaneous generation. An important feature of his work was that he gave a theoretical analysis of this phenomenon and founded his theory of spontaneous generation. In the course of time it seems that his views changed, but in the last analysis they served as the basis of the idealistic hypotheses concerning the origin of life.

Aristotle considered that living things, like all other concrete objects (substances), are formed by the conjunction of some passive principle, 'matter' (by this word Aristotle obviously meant what we now call material), with the active principle of 'form'. The 'form' of living things manifests itself in the 'entelechy of the body' the soul. This shapes the body and sets it in motion. Thus matter does not possess life but is infused with it. It is adapted and organised by means of a spiritual force; an orientating internal substance (entelechy) brings matter to life and sustains the living thing. The spirit, however, is already inherent in the actual elements from which living things are formed, it is inherent in a smaller degree in the earth and in a greater degree in water, air and fire. Because of this, that which is created by the spirit depends substantially on the preponderance of this or that element. Earth produces mainly plants; water, aquatic animals; air, the inhabitants of the land; and fire, the supposed inhabitants of the celestial bodies, in particular, the moon. For their 'form' living things which arise from others like themselves depend on 'animal warmth' and when they arise by spontaneous generation on 'solar warmth'. Thus, in spontaneous generation, decaying materials do not on their own give rise to life; they are brought to life under the influence of the light of the sun which gives 'psychic warmth'.

The views of Aristotle exerted an enormous influence on the whole subsequent history of the problem of the origin of life. Aristotle, with his undisputed authority, supported the results of direct, naive observation, and for many centuries ahead prejudiced further study of spontaneous generation. All the later philosophical schools, both Greek and Roman, completely shared the opinion of Aristotle on the possibility of spontaneous generation of living beings. Moreover, as time went on the theoretical basis of the 'phenomenon' took on a more and more idealistic, indeed even mystical, character.

A whole series of writings, from the 3rd and 2nd centuries B.C., contain numerous tales and 'miraculous stories' of plagues of lice' in which the juices of the human body are changed into parasites, of the appearance of worms and insects from rotting materials, of crocodiles from the mud of the Nile, and so forth. Concerning such matters, the most authoritative philosophical school of that time, the Stoics, taught that animals and plants originate as a result of the activity of 'engendering force' which is a property of *pneuma*.

From the later Stoics this view obtained a wide circulation in both East and West, through a number of philosophers and writers, particularly the much-travelled Poseidonius. It thus obtained general recognition at the beginning of our own era. In scientific treatises, in political pronouncements and in artistic productions of that period we meet continually with descriptions of various cases of spontaneous generation. We find them in the works of Cicero, of the famous geographer Strabo, of the versatile scholar Philo of Alexandria, of the historian Diodorus Siculus, of such poets as Virgil and Ovid, as well as in the works of the later writers Seneca, Pliny, Plutarch and Apuleius.

The idealistic character of the teachings concerning spontaneous generation was clearly expounded by the neo- Platonists (in the third century A.D.). The leader of this philosophical school, Plotinus, taught that living things could originate from earth, and that this method of origin was not confined to the past but also continues now, in the course of decay. He explained this phenomenon as the result of the animation of matter by a life-giving (*vivere facit*) spirit, and it seems that he was the first to formulate the concept of the 'Life Force' which has persisted even up to the present in the teachings of the contemporary vitalists.

Early Christianity borrowed guiding ideas concerning spontaneous generation from the *Bible*, which in its turn borrowed its material from the mystical tales of Egypt and Babylon. The theological authorities of the end of the fourth and beginning of the fifth centuries A.D., 'the fathers of the Christian Church', combined these legends with the teachings of the neo-Platonists and elaborated their mystical conception of the origin of life on this basis.

Living in the middle of the fourth century A.D. was St. Basil the Great who was then and still is one of the leading religious authorities of the Eastern Church. It was under his influence that the leaders of the Orthodoxy formulated their beliefs concerning the origin of life. His book *Hexaëmeron* still retains its place in Church literature, particularly in the Russian language. Discussing the problem in which we are interested, he writes as follows:

> *For if there are creatures which are successively produced by their predecessors, there are others that, even today, we see born from the earth itself. In wet weather she brings forth grasshoppers and an immense number of insects which fly in the air and have no names because they are so small; she also produces mice and frogs. In the environs of Thebes in Egypt, after abundant rain in hot weather, the country is covered with field mice. We see mud alone produce eels; they*

do not proceed from an egg, nor in any other manner; it is the earth alone which gives them birth.

According to Basil the Great all these instances of the spontaneous generation of life (many of which were obviously borrowed from Aristotle) occurred by divine command which has continued to act with undiminished force from the creation of the world to the present day.

St. Augustine of Hippo is a high authority for the Western Church like St. Basil for the Eastern. He also accepted the spontaneous generation of living things as an unchanging truth and strove in his teachings simply to bring the phenomenon into line with the world philosophy of the Christian Church. Similarly, he wrote "God as a rule creates wine from water and earth through the mediation of grapes and their juice; however sometimes, as in Cana of Galilee, he can create it directly from water. Thus also, in respect of living things, he may cause them to be born from seeds or to emerge from inanimate matter where invisible spiritual seeds (*occulta semina*) repose."

Thus Augustine saw in the spontaneous generation of living things a manifestation of divine will the animation of inert matter by the 'life-creating spirit'. In this he affirmed a doctrine concerning spontaneous generation which was in complete agreement with the dogmas of the Christian Church.

Throughout the Middle Ages a belief in spontaneous generation held undivided sway over people's minds. Mediaeval philosophical thought could exist only as theological thought, embodied in one or another doctrine of the Church. Any kind of philosophical question could only obtain a hearing if it was linked with one or another theological problem. Philosophy became the 'handmaid of theology', *ancilla theologiae*. The problems of science were relegated to a lower plane. People did not use observation and experiment as a guide to an understanding of nature but used instead the teachings of the *Bible* and of theological treatises. Only a very scanty knowledge of the problems of mathematics, astronomy and medicine penetrated into Europe from the Arab and Hebrew teachers.

It was in this way that the works of Aristotle first reached the European peoples, though often in the form of garbled translations. At first his teachings appeared dangerous, but later, when the Church appreciated the full usefulness of these teachings for many of its purposes, it raised Aristotle to the status of 'the forerunner of Christ in the realm of nature' (*praecursor Christi in rebus naturalibus*). Accordingly, in the apposite words of V. Lenin, "the scholasts and clerics seized upon that which was dead in Aristotle and not upon that which was alive". This teaching was widely accepted by theologians in the Middle Ages, especially insofar as it concerned the origin of life. They held that the animation of lifeless matter by the 'eternal divine spirit' constituted the essence of it.

As an example one may here quote from one of the greatest exponents of scholastic Aristotelianism, the Dominican Albert von Bollstädt, known as Albertus Magnus (1193-1280). According to tradition, Albertus Magnus took a great interest in zoology, botany, alchemy and mineralogy. But in his numerous works he assigns considerably less place to independent observations than to material borrowed by him from ancient authors. On the question of the origin of life Albertus Magnus consistently supported the theory of spontaneous generation, and in his book *De mineralibus* he specially emphasised the fact that the origin of living things in the presence of decay occurs as a result of the 'animating force' (*virtus vivificativa*) of the stars.

In his writing on zoology Albertus Magnus gives many accounts of the spontaneous generation of insects, worms, eels, mice, etc., from various sorts of decaying materials, from moist earth, vapours, sweat and various forms of filth. In just the same way vapours of the earth and water give rise, under the influence of warmth and the light of the stars, to numerous plants, not only fungi but even to herbs, bushes and trees which often grow in places where their seed cannot have been carried.

The pupil of Albertus Magnus, Thomas Aquinas (1225-1274) also held such opinions. In his chief work, Summa Theologica, he deals with questions concerning the origin of life. In doing so he relies partly on the views which he ascribed to Aristotle and partly on the teachings of Augustine about the 'anima vegetativa'. He thus freely accepted the possibility of the spontaneous generation of such animals as, for example, worms, frogs and snakes as an effect of the warmth of the sun in the presence of decay. Even those worms which torment sinners in the infernal regions arise, according to the opinion of Thomas Aquinas, in this way from the rotting of their sins. In general Thomas believed in and preached a militant demonology.

He taught that the Devil really exists as the chief of a whole horde of demons. Hence he conceived the idea that various forms of pest harmful to man can arise as the result of tricks of the Devil and the spirits of evil subservient to him.

The practical results of this hypothesis manifested themselves in the numerous trials of witches who were charged with letting loose mice and other pests on to the fields and thus destroying the sown seed. And it is well known that Catholic bishops also used all sorts of spells and exorcisms in an effort to cast out worms, mice, cockchafers and other harmful creatures from the fields of those who had been confided to their care. According to Uhland, the Swiss and Tyrolese bishops in the sixteenth century laid the curses of the Church on all sorts of agricultural pests and, according to Bodenheimer, ceremonies of this sort persisted until the end of the eighteenth century.

We have dwelt at some length on the views of Thomas Aquinas because, to this day, his teaching is acknowledged by the Catholic Church as the only true philosophy. Thus, the Western Church has retained through all the centuries the principle of the spontaneous generation of living things according to which living things originate from inanimate matter as a result of animation by a spiritual principle.

The standpoint of the theological authorities of the Eastern Churches is similar. In this matter they rely chiefly on the pronouncements of Basil the Great. The opinions on this subject of the outstanding and active participants in the work of the Russian Church, Dimitrii Rostovskii and Theofan Prokopovich, though formulated as late as the eighteenth century, may serve as an illustration. Dimitrii, bishop of Rostov, lived in the time of Peter I and in his works *Annals relating shortly the acts from the beginning of the world until the birth of Christ* (1708) he wrote that Noah did not take in his ark those animals which are capable of spontaneous generation; they were destroyed on the ground by the flood and then arose anew.

Moreover, from the moisture of the earth, from decay and putrefaction, there arise mice, toads, scorpions and other creatures which creep upon the earth, and various worms and even beetles, cockchafers and cockroaches; and also from heavenly dew there are conceived midges and gnats and other such things. These all perished in the Flood and after the flood they arose anew from such beginnings.

In the course of theology which he gave in the Ecclesiastical Academy in Kiev Theofan Prokopovich developed, almost word for word, the same idea.

Furthermore, there is a multitude of animals which arise without copulation of the parents; independently, from rotten things, and there was thus no necessity to give shelter in the ark to creatures such as mice, worms, wasps, bees, flies and scorpions.

Even in the nineteenth century a translation of a book by W. Frantze was published by Benjamin, archbishop of Nizhegorod, in which it was stated that insects, worms, frogs and mice arise by spontaneous generation "from rotting tree stumps, from the dung of animals, from the sand of the sea, from decaying earth, from corpses etc."

As we have already pointed out, science was at a very low ebb in mediaeval Europe. It was in complete subjection to theology. The natural phenomena observed by the travellers and learned men of those times were not only discussed, but also described, as though scholastic wisdom demanded that they should be in complete conformity with the Church dogmas. The works of the learned men of the Middle Ages therefore abound in those same fantastic descriptions and sometimes even sketches of the spontaneous generation of various insects, worms and fishes from slime and damp earth, of frogs from the dews of May, and even of lions from the stones of the desert. It is specially characteristic of the mediaeval methods of the study of nature that at this time there was a wide diffusion of lore concerning goose trees, vegetable lambs and homunculi.

According to the testimony of very authoritative men of learning of those times, geese and ducks arise from barnacles which in their turn are derived from the fruits of trees. From these latter, birds may also be formed directly.

We find this tale of the goose tree as early as the beginning of the eleventh century in the works of Cardinal Peter Damian (1007-1072). The English encyclopaedist Alexander Neckam (1157-1217) considered that birds are formed from the resin of conifers on contact with the salt water of the sea. Furthermore, this story of the vegetable origin of ducks and geese became so widely accepted that their meat was used as lenten fare though this was later forbidden by a special order of Pope Innocent III (1198-1216).

But in spite of this, almost three centuries later, at the end of the fifteenth century, the nobleman Leo von Rozmital described a dinner given in his honour in London by the Duke of Clarence at which, as a hot dish described as fish (for lenten fare), were served ducks, which there generate themselves from 'worms' in the sea. However, Rozmital remarks that the taste of these 'fish' was exactly like that of ducks.

It is interesting that the story of the goose tree persisted until the end of the sixteenth and even the beginning of the seventeenth century. A series of authors describe their personal observations on this subject and even give more or less fantastic drawings showing how the birds are gradually formed from the fruits of the tree.

Evidently this legend was based on the naive interpretation of superficial observations of barnacles of a special kind. In the adult state these marine animals attach themselves by a special kind of stalk to rocks, stones, the bottoms of ships and trees which have accidentally fallen into the water. On the shores of the north of Scotland, Ireland and the neighbouring islands this happens at the time when flocks of young Arctic geese fly there from the north.

These two phenomena were confused and fantasy, not knowing where they came from, drew a picture of the formation of birds from the barnacles found on the branches of trees. It may also be that analogous superficial observations formed the basis for the other stories concerning vegetable lambs. The well-known traveller Odoric di Pordenone (d. 1331) was the first to record this. It was related to him by 'reliable' people that in the Tatar kingdom of Khadli there grew enormous gourds which opened when they were ripe to reveal within themselves lambs covered with white wool and having very delicious meat. 'Sir John Mandeville' described his travels in Eastern lands and also told stories of a whole tree, from the melon-shaped fruits of which there arose living sheep. This story persisted for centuries and as late as the middle of the seventeenth century it was repeated anew by Adam Olearius in his descriptions of his travels in Muscovy and Persia. He wrote:

> *We were told that there beyond Samara, between the rivers Volga and Don, there grows a rare form of melon or rather pumpkin which*

is very like an ordinary melon in size and shape but its appearance reminds one of a lamb because it has clearly defined limbs. The Russians therefore call it the 'little ram'. This 'vegetable lamb' feeds on the grass around it but frequently falls a prey to wolves, which are very fond of it.

Later, Olearius writes that he had the good fortune actually to see the wool of such a sheep. The story of the homunculus developed on the basis of alchemical experiments. It is known to have made its appearance as early as the first century A.D. This story was based on the supposition that by mixing the passive maternal original substance with the active masculine one it is possible to reproduce artificially the phenomenon of birth and to obtain the embryo of a tiny person *homunculus*.

Like the legend of the goose tree and the vegetable lamb, stories about the homunculus were current throughout the Middle Ages and are to be met with in many alchemical treatises. A typical exponent of the earlier natural philosophy of the sixteenth century, Theophrastus Bombast von Hohenheim, known as Paracelsus (1498-1541), even gives an 'exact receipt' for the preparation of homunculi. For this it is necessary to obtain human sperm, place it in a sealed gourd inside a horse's stomach and during the course of a certain time to carry out a series of complicated manipulations. In this way there is formed a small person complete in all its parts, like children born of women but on a far smaller scale.

In general, Paracelsus was a convinced supporter of the spontaneous generation of living things. He maintained that there is an active life force, the *arche*, which governs the bodies of animals and men and which can be controlled by means of magic remedies. This force itself determines the formation of the organism and its later conduct. Paracelsus developed a theory of spontaneous generation of life with this philosophical outlook. He even produced a number of personal observations of the sudden formation of mice, frogs, eels and tortoises from water, air, straw, rotten wood and all sorts of rubbish. The descriptions of the views and beliefs of the learned men of the Middle Ages were excellently portrayed in Goethe's tragedy *Faust*. Here Mephistopheles refers to himself as *"Der Herr der Ratten und der Mäuse, der Fliegen, Frösche, Wanzen, Läuse"*, and a swarm of insects fly out from his old doctor's fur cloak and praise him not only as their patron but also as their father, as though he had actually begotten them there and then.

The part played by the homunculus in the second part of *Faust* is also well known. Wagner takes great pains with the preparation of his alchemical experiments. For this he mixes hundreds of substances, corks them up in a retort and proceeds to purify them by distillation. If the conjunction of the stars were favourable a manikin should develop in the retort. But even in this case the spontaneous generation did not occur without the intervention of Mephistopheles, whom the homunculus greeted as his 'cousin'.

In the second half of the sixteenth century and, in particular, in the seventeenth century, observations of natural phenomena were getting more accurate. Copernicus (1473-1543), Bruno (1548-1600) and Galileo (1564-1642) destroyed the old Ptolemaic system and drew up sound theories concerning the universe of stars and planets which surround US. However, this blossoming of exact knowledge did not as yet touch upon biological problems. The idea of the primary spontaneous generation of living things remained unchallenged in the minds of the investigators of that time.

As an example we may here mention the well-known physician of Brussels, van Helmont (1577-1644). He used some methods of exact experiment which enabled him to make substantial progress in the study of the complicated problem of the nutrition of plants. Nevertheless, he was quite convinced that living things could arise by spontaneous generation and even went further and carried out a number of observations and experiments to confirm the hypothesis. For example, he gives a well-known receipt for making mice from grains of wheat. He held that human sweat could serve as the life-giving principle. For this it was necessary to place a dirty chemise in some sort of receptacle which contained wheat grains. After 21 days the 'fermentation' was stopped and the exhalations from the shirt together with those of the corn had formed living mice. It was especially surprising to van Helmont that these artificially produced mice were exactly like those born from the seed of their parents.

Neither did Harvey (1578-1657), the originator of the theory of the circulation of the blood, reject the idea of spontaneous generation. However, although the celebrated phrase *omne vivum ex ovo* (everything alive comes from an egg) belongs to him, he was here giving a very wide meaning to the word egg. He considered *generatio aequivoca* (spontaneous generation) of worms, insects, etc., to be perfectly possible as a result of the activity of special forces which develop during putrefaction and similar processes.

This also was the view of Harvey's contemporary, the founder of seventeenth century English materialism, Francis Bacon (1561-1626). In his works he expressed the opinion that various plants and animals (such as flies, ants and frogs) could arise spontaneously in the course of the decay of various materials. However, he approached this phenomenon from a materialist position and saw in it only a proof of the absence of an impassable barrier between the inorganic and the organic world.

The materialistic interpretation of spontaneous generation was particularly clearly expressed by Descartes (1596-1650). This great French philosopher, although he believed the spontaneous development of living things to be beyond dispute, nevertheless categorically denied that this emergence occurred under the influence of the *anima vegetativa* of the scholasts, the *arche* of Paracelsus, the 'spirit of life' of van Helmont or any other spiritual principle. In sharp contradistinction to the religious teachings then prevailing and to the anthropocentric tendencies of mediaeval natural philosophy, Descartes tried to relate the qualitative diversity of natural phenomena to matter and its movement.

Thus, according to Descartes, the living organism does not need to be explained by any special obedience to 'a vital force'. Descartes postulates nothing other than a machine, very complicated certainly, but of completely intelligible construction, whose movements depend exclusively on the pressures and interactions of particles of matter as do the movements of the wheels in a clock. Thus different kinds of living beings can arise spontaneously from the surrounding lifeless matter. In particular, when moist earth is exposed to the rays of the sun or when putrefaction occurs, there develop all kinds of plants and animals such as worms, flies and a variety of insects. But for this to happen there is no need for any intervention whatsoever by any 'spiritual principle'. Spontaneous generation consists only of the natural process of self-formation of complicated machines, a process which takes place invariably when certain circumstances, not yet fully investigated by us, are fulfilled.

Thus, down to the middle of the seventeenth century, the actual possibility of spontaneous generation had not been seriously questioned by anyone. The dispute between the mystical doctrines from the Middle Ages and the materialism now in violent spate was only concerned with the theoretical treatment of the phenomenon: was spontaneous generation to be regarded as a manifestation of 'a spiritual principle' or as a natural process of self-formation of living beings? However, the study of living nature was all the time becoming both wider and more accurate in its approach, and the assurance of those who had accepted spontaneous generation as a 'fact' now began to be shaken.

Redi's Experiments

In this matter the experiments of the Tuscan physician Francesco Redi (1626-1697) can justly be counted as the turning point. To Redi fell the honour of being the first to emerge with the support of experiment from the belief in spontaneous generation which had ruled without interruption for so many centuries. In his treatise *Esperienze intorno alla generazione degl' insetti* (1668) he describes a series of his experiments which show that the white maggots in meat are simply the larvae of flies. He kept meat or fish in a large vessel, covered with the finest Neapolitan muslin, and, for still more complete protection, covered the vessel with a frame on which muslin was stretched. Although plenty of flies alighted on the muslin, no maggots appeared in the meat. Redi pointed out that he had succeeded in observing how the flies laid their eggs on the muslin, but that only when these eggs fell on to the meat did they develop into meat maggots. From this he concluded that decaying substances are only a place or a nest for the development of insects, but that the laying of eggs is an essential preliminary to their development; without eggs the maggots never appear.

It should not be thought, however, that Redi had succeeded in completely ridding himself of the notion of spontaneous generation. In spite of his brilliant experiments, which he had interpreted correctly, this learned man freely admitted the possibility that spontaneous generation might occur in other cases. Thus he states that worms in the intestines or in timber arise on their own from rotting materials. Moreover, in his opinion, the maggots which are found in oak galls are formed from the juices of the plant. Only later was this opinion refuted by the investigations of the scientific physician Vallisneri (1661-1730).

This example makes it clear that what has been repeated for centuries (though often wrongly) is not easily confuted. Throughout the eighteenth century, and even in the beginning of the nineteenth century, many scientists and philosophers of different tendencies and schools, and even more writers and poets, often described in their works various fantastic instances of the spontaneous generation of beasts, fishes, insects and worms, or made it clear that they considered that such a phenomenon was quite possible.

As observations of nature became more refined and, in particular, knowledge of the structure of living things became more detailed, so it was admitted, though only very gradually, that the spontaneous generation of such complicated things from structureless filth and decaying matter was impossible. In this way the belief in the spontaneous generation of all the more highly organised things ceased to be held among scientists. But this idea as to the primary origin of living things did not disappear. On the contrary, during the eighteenth and nineteenth centuries it reached its fullest development in connection with the simplest living things, the micro-organisms.

Hypotheses Concerning the Spontaneous Generation of Microbes

Almost at the same time as Redi was carrying out his celebrated experiments, a new world of living creatures invisible to the naked eye was opened up by the Dutch scientist Anthony van Leeuwenhoek (1632-1723), with the help of magnifying glasses made with his own hands. In letters to the Royal Society in London he described in detail these small 'living animalcules' discovered by him in rain water which had stood for a long time in the air, in various infusions, in excrement, in the tartar of teeth, etc. With his glass van Leeuwenhoek saw representatives of almost all the classes of micro-organism known to us at the present day. He gave descriptions, which were surprisingly accurate for those times, of infusoria, yeasts, bacteria, etc.

The curious discoveries of the Dutch scientist attracted the most general attention and provoked many similar studies. Micro-organisms were discovered wherever decay or fermentation of organic substances was going on. They were found in different sorts of plant infusions and decoctions, in decaying meat, in stale broth, in sour milk, in fermenting wort etc. Substances which quickly become tainted or which decay easily had only to be kept in a warm place for some time when microscopic living things, which had not been there before, at once began to develop in them. As the belief in the spontaneous generation of living things was current at the time, it was unhesitatingly assumed that it extended to cover the spontaneous generation of living microbes from inanimate matter in these decoctions and infusions.

Van Leeuwenhoek himself did not propose this idea. He maintained that the micro-organisms fell into his infusions from the air. This opinion was confirmed by the experiments of Louis Joblot. This distinguished follower of van Leeuwenhoek used infusions of hay which were swarming with micro-organisms, boiled them for 15 minutes and then poured equal parts into two vessels. One of these he covered closely with parchment before it cooled, the other was allowed to stand uncovered. In the open vessel very small living things (apparently infusoria) grew abundantly, but they did not appear in the closed one. At the end of the experiment the parchment was removed from the closed vessel too, after which the infusion was soon populated with micro-organisms. However, the experiments of Joblot were not convincing enough for his contemporaries and were later completely forgotten.

Philosophical thought at that time could still not renounce the principle of spontaneous generation and, as before, the dispute between the different schools was concerned not with whether or not microbes can develop of their own accord, but only with the spiritual or material basis of this apparently self-evident 'phenomenon'.

The discovery of the extremely small germs of life which were to be found everywhere was expressed in the philosophical system of G. Leibnitz (1646-1716). His teachings about monads included metaphysical rehashing of the contemporary data of mathematics and science. According to Leibnitz the monads are primary centres of spiritual force. As the ultimate sources of everything they must be characterised by absolute simplicity and individuality. Matter being inherently passive, the monads constitute the spiritual substance, for only the spirit, in Leibnitz's view, has the capacity for uninterrupted activity.

Starting from these assumptions, Leibnitz considered that life cannot be explained simply on the basis of bodily forces. In particular, he considered the possibility that higher plants and animals could arise by spontaneous generation from decaying material as disproved by direct experiment. The development and disappearance of living things is but the evolution and involution of eternally existing germs. Those substances which we usually consider inorganic contain within themselves a whole world of germs of life. "Even in vinegar and bookbinder's paste," wrote Leibnitz, "these germs are present." Thus, all bodies can contain within themselves organic structures, but these are still invisible, incomplete, and only in the form of germs. In these germs there are already present and pre-existing all the conditions for future specific organisation. Thus, living things are formed spontaneously from them by later development.

We find the same ideas concerning spontaneous generation in the works of the French scientist G. L. Buffon (1707-1788). He also considered that the whole of nature is full of 'ubiquitous units or germs of life' but, in opposition to Leibnitz, he attributed to them a material character. These material particles endowed with life are capable, according to Buffon, of uniting with one another to form lower plants and animals from which the highly organised creatures later evolve. Conversely, on the decay of the body, individual existence ceases but living particles of matter which were at first scattered and then entered into its composition can now, once more, unite into living bodies. From them microbes originate. In this Buffon saw the explanation of the phenomenon of the spontaneous generation of microscopic organisms in putrefying organic liquids and infusions.

This view was shared by the contemporary and friend of Buffon, the Welsh Roman Catholic priest and naturalist J. T. Needham (1713-1781). He believed that in each microscopic particle of organic matter there was concealed a special 'vital force' which could animate the organic matter in an infusion. Thus Needham developed vitalistic views, which were very common in those days, concerning the essence of life and its begetting. However, Needham's importance in connection with the problem which we are considering depends, not only on his views, but also on the extensive experiments which he carried out in an effort to confirm the spontaneous generation of micro-organisms. He says:

> *I took a quantity of mutton gravy hot from the fire and shut it up in a phial closed with a cork so well masticated that my precautions amounted to as much as if I had sealed my phial hermetically. I thus excluded the exterior air that it might not be said my moving bodies drew their origin from insects or eggs floating in the atmosphere. I neglected no precaution even so far as to heat violently in hot ashes the body of the phial that if anything existed even in that little portion of air which filled up the neck it might be destroyed and lose its productive faculty.*

But, in spite of all this, after some days the vessel swarmed with micro-organisms. He made similar investigations on a variety of organic liquors and infusions, always with the same result. This naturally led him to the conclusion that it was completely possible, and indeed inevitable, for micro-organisms to arise spontaneously from putrefying organic substances.

However, these experiments of Needham were subjected to severe criticism by an Italian scientist, the priest Spallanzani (1765). Spallanzani, like Needham, carried out experiments with the object of establishing or refuting the possibility of spontaneous generation, but, on the basis of these experiments, he arrived at exactly the opposite conclusion. He asserted that the experiments of Needham had succeeded because of insufficient heating of the vessels containing the liquid, resulting in their inadequate sterilisation. Spallanzani himself carried out hundreds of experiments in which plant decoctions and other organic liquids were subjected to more or less prolonged boiling, after which the vessel containing them was sealed and thus the access of air to the liquids was prevented. Air, according to Spallanzani, carried the germs of micro-organisms. Whenever the operation was conducted with proper attention the liquids contained in the vessel did not putrefy and living creatures did not appear in them.

Needham objected to this that on prolonged heating of the liquids the air contained in the vessels was spoilt and that this was the chief reason for the failure of microorganisms to develop. Secondly, he asserted that on prolonged heating the 'vital force' of the organic infusions was destroyed. This 'vital force' usually seems to be capricious and inconstant and cannot withstand prolonged and severe treatments. Thus Needham considered, not that he had heated the liquids too weakly but, on the contrary, that in the experiments of Spallanzani these liquors had been heated too strongly and the generative power of the infusions had thus been destroyed.

In order to refute this Spallanzani carried out fresh experiments. In a long series of tests conducted with exceptional care he answered nearly all the criticisms that had been made by Needham. Nevertheless, he did not succeed in convincing his contemporaries and the controversy remained unsettled for very nearly a hundred years longer.

It is interesting to note that, in parallel with Spallanzani, in the period between the publication of his first and second works, analogous experiments were being carried out by the Russian M. Terekhovskii, who was sent from St. Petersburg to Strasbourg for scientific investigations.

In his dissertation, De chao infusorio Linnaei, which he published in 1775 in Latin, Terekhovskii recorded the results of his extensive investigations on the 'animalcules of liquors', i.e. the microscopic living creatures which appear in all kinds of organic infusions the infusoria, flagellates and other primitive organisms. In his opinion it was absurd to suppose that even the very simplest organisms with all the extraordinary complication of their structures which "no mechanic, even the most skilful who ever lived, could understand completely, try as he might, still less reproduce" might "be formed by chance from a chaotic mixture of inanimate particles". In effect, as S. Sobol' pointed out, the numerous and very carefully performed experiments of Terekhovskii showed that "the spontaneous generation of animalcules does not take place under any conditions". However, these statements and experiments of the Russian scientist, which we now know were completely correct, did not receive recognition in the scientific world of that time and were quickly forgotten.

The doctrine of spontaneous generation was still defended by many scientists and philosophers in the end of the eighteenth century and beginning of the nineteenth century. In particular, it was developed by representatives of the German idealistic philosophy. I. Kant (1724-1804) himself considered that the primary internal cause of the development of organisms was supernatural (metaphysical) and that therefore the hypothesis of spontaneous generation was merely a 'bold adventure of the intellect'. However, the later *Naturphilosophen*, G. Hegel (1770-1831), F. Schelling (1775-1854) and L. Oken (1779-1851) extensively developed the idea of *generatio aequivoca*. Thus, for example, Hegel stated that the earth and the sea had a clear need to be vivified "but in its general form, vivification seems to be *generatio aequivoca*"; and further, in his *Enzyklopädie* he wrote that "the earth and, in particular, the sea generate all sorts of lichens, infusoria, innumerable phosphorescent living specks".

According to Schelling, there is a complete identity between the earth and the animal and plant world. The earth itself is transformed into plants and animals because that which is called dead matter is merely the 'dormant animal and plant world'.

Oken, who was a follower of Schelling, developed the idea that the earth, in the course of its metamorphosis, degenerates into carbon and that this, being mixed with water and air, is converted into 'hydrated oxidised carbon' which, as a formless primaeval slime, acts as the basis of all organisms which have a form. Every living thing arises from this slime. At first, like the primaeval planets, it turns into spherical globules (the globules of primaeval slime) or infusoria under the influence of light. These later metamorphose into plants and animals which afterwards, on putrefaction, give rise again to infusoria. Moreover, it is also possible that spontaneous generation of ticks, worms and such creatures occurs by simple direct coagulation of the primaeval slime.

Thus, we find in the works of Oken, along with a banal conception of the spontaneous generation of life, the elements of a specifically scientific prediction. He had already put forward the theory of the development of life by the gradual evolution of matter, although in a very confused form.

While these discussions on natural philosophy were taking place in the first half of the nineteenth century, a whole series of experiments was carried out with the aim of establishing or refuting the possibility of the spontaneous generation of microbes.

An exceptional amount of care and experimental skill was expended on elucidating the significance of air in the appearance of living things in liquids which had been previously heated.

The well-known French chemist J. L. Gay-Lussac (1778-1850) showed, by means of direct analyses, that oxygen, that is the component of the air which sustains burning and breathing, is absent from vessels containing liquid which had been sealed up after boiling. This confirmed Needham's view. To elucidate the part played by oxygen, Gay-Lussac filled with mercury a glass tube which was closed at one end (a eudiometer) and stood it in a vessel of mercury with the closed end uppermost. A grape was then inserted under the mercury into the tube and crushed with a wire which was introduced through the mercury. The juice which ran out of the grape occupied the upper part of the tube. It remained transparent and apparently completely sterile for a long time. However, after the admission of a bubble of air, the juice quickly began to ferment and to be inhabited by microorganisms.

This experiment, which was later made great use of by the adherents of spontaneous generation, is interesting from the point of view that in it the source of infection was, as we know now, the germs of the micro-organisms which were present on the surface of the mercury, to which neither the experimenter himself nor any of his later interpreters had paid any attention.

The German naturalist T. Schwann made a new test of the significance of oxygen for the spontaneous generation of microbes. He caused a stream of heated air to pass through a glass tube into a vessel containing sterile meat broth and showed that in these circumstances the broth did not putrefy. Hence spontaneous generation did not proceed in the presence of a constantly renewed stream of sterilised air. However, a repetition of this experiment using a liquid containing sugar gave completely different results. In spite of the fact that, according to the author, the methods used in them were exactly the same as those used in the experiments with the broth, a mass of living micro-organisms often developed.

In the same year F. Schulze carried out analogous experiments differing only in that the air which was admitted into the vessel with the sterilised liquid was freed from germs, not by heating but by being passed through strong sulphuric acid. The results were the same. However, numerous repetitions of Schulze's experiments gave inconsistent results and in some cases micro-organisms appeared in the liquids. This, as we now know, depended on the invasion of the liquid by spores which were present in a resistant state in the bubbles of air passing through the sulphuric acid.

A little later (1853) the Heidelberg professors H. Schroder and T. Dusch simplified the experiment still further by purifying the air by passing it through a layer of sterilised cotton wool which served as an excellent filter, removing all germs of micro-organisms. Thus they were able to free the air from germs while not submitting it to any chemical treatment or applying heat to it. In fact, a series of experiments was made by these workers with meat broths, and the wort of beer. These were boiled and then allowed to stand for many weeks without any change occurring. However, milk and meat without water went bad quickly under these conditions and became full of micro-organisms.

Although all the experiments which had been carried out tended to refute the possibility of spontaneous generation, their evidence was not strong enough, in that they were sometimes unsuccessful for no demonstrable reason and microorganisms appeared in the liquid. We now know that this occurred as a result of the accidental introduction of organisms owing to some technical fault; however, contemporary scientists did not see the matter in that light. All these failures, in spite of a known wish to succeed, might easily be interpreted, and were in fact interpreted, as indicating that spontaneous generation, though not universal, could take place under certain circumstances. This opinion was held even by such outstanding investigators as Dumas, Naegeli and a number of other scientists of the middle of the nineteenth century.

The conflict of opinion concerning the possibility of the spontaneous generation of micro-organisms attained its greatest naiveté in 1859 when F. Pouchet published a paper in which he tried to prove this possibility experimentally. In his voluminous work, comprising about 700 pages, Pouchet developed his theory of spontaneous generation, which is fundamentally very reminiscent of the views of Needham. Fermentation or decay of organic substances precedes each manifestation of spontaneous generation. Only substances forming part of living organisms can give rise to new life. Under the influence of fermentation or decay the organic particles of the corpse disintegrate but, having wandered around for some time independently, they become united once more by virtue of their inherent properties and thus new living things are created. Pouchet considered that a 'life force' was a prerequisite for the development of living things and therefore he never believed that living things could arise de novo in mixtures of mineral substances. In confirmation of his views Pouchet made a large series of experiments in which he repeated the investigations of his predecessors. In these he always got results in agreement with his own ideas; that is to say, micro-organisms always developed in his organic liquids.

Only about a hundred years separate us from the experiments of Pouchet, but when one reads about these experiments now one cannot help noticing how crudely and messily they were carried out. Pouchet, for example, categorically denied the possibility that germs of micro-organisms might have got into his infusions and solutions from outside simply because "Joly and Musset carried out careful chemical analyses of the surrounding air". But what could they find out in this way even if thousands of bacteria and spores were hovering around them? In just the same way Pouchet asserted, without any foundation, that his original hay infusions certainly did not contain the germs of any microorganisms. However, we know that enormous numbers of such germs are always present on the surface of hay and that, on simple infusion of the hay with water, which is what Pouchet did, these germs must certainly fall off into the infusion in a perfectly viable state. This clearly occurred, for when Pouchet placed his hay infusions in a warm place for six days there appeared in them not only bacteria, but also such highly organised creatures as infusoria, in the cells of which there are digestive vacuoles, mouths and other very complicated and specialised organs. It is quite clear to us now that under such experimental conditions the appearance of infusoria was simply due to their germs always having reached the original solution from the surface of the hay. This may easily be demonstrated nowadays by direct observation. Pouchet's statement that spontaneous generation of infusoria occurred in his infusions sounds quite unjustified and even ridiculous in the light of present-day knowledge. However, Pouchet's work made a great impression on his contemporaries.

Work of Pasteur

The French Academy of Sciences awarded a prize to whoever, by means of accurate and convincing experiments, should cast light on the question of the primary origin of living creatures. This prize was awarded to Louis Pasteur who, in 1862, published his work on spontaneous generation in which, by a series of conclusive experiments, he demonstrated the impossibility of the formation of micro-organisms from various infusions and solutions of organic substances. Pasteur was successful in doing this only because he left the beaten track of blind empiricism and approached the whole problem broadly in his experiments. He also gave a rational analysis of all earlier experiments and explained the mistakes of those who carried them out. First of all Pasteur cleared up the question of the presence of micro-organisms in the air which, as we have seen above, was considered to be one of their chief origins. The partisans of spontaneous generation, Pouchet in particular, repeatedly expressed doubts as to whether germs of life were really present in air and demanded a demonstration of the 'infinite mass of microorganisms' which are present in the air.

Pasteur solved this problem by a very simple method. Using an aspirator he drew air through a tube into which a plug of gun cotton had been inserted. As Schröder and Dusch had already shown, all the smallest particles are retained by the cotton and remain in the tube. The current of air was maintained for 24 hours and the plug with the dust which had been caught in it was removed and dissolved in a mixture of alcohol and ether. At this stage all the solid particles present sank to the bottom. They were washed with solvent and then studied under the microscope. There were always found thousands of organised bodies which differed in no way from the common micro-organisms and their spores. The presence of large numbers of organised bodies in the ambient atmosphere had thus been demonstrated.

Furthermore, Pasteur showed that these germs which are present in the air can often initiate the growth of organisms. First of all he repeated the experiments of Schwann with some variations and improvements. The boiling of the organic liquids was carried out in a round-bottomed flask with a long drawn-out neck joined to a platinum tube which was heated to red heat with a gas burner. Thus, the air which was drawn into the flask when the liquid in it had finished boiling passed through a red-hot platinum tube in which all the germs present in it were sure to be destroyed. While passing from the tube to the flask the air was cooled by a stream of water. After it had been filled with air the flask was sealed and in this state it could be kept indefinitely. When the experiment was set up in this way the liquid never decomposed and no micro-organisms were formed. However, if the sealed neck of the flask was broken and a cotton plug through which air had been passed was thrown into the liquid contained in it and the neck was quickly sealed again, then the liquid soon became filled with moulds, bacteria and even infusoria. This meant that the liquid had not lost its nutrient capacity for micro-organisms and the germs which had been present in the air and were collected on the cotton plug could, in fact, easily develop in such liquids.

Later Pasteur sterilised the air admitted to the flask without heating it. For this purpose he relied partly on the method of Schröder and Dusch, drawing the air through a cotton-wool plug, and partly brought his own native skill to bear on it. As usual, Pasteur half filled the roundbottomed flask with the experimental liquid and then softened the neck of the flask in a flame and drew it out. The part which was drawn out was bent into the shape of the letter S. The contents of the flask were then boiled without any further precautions. When a strong current of steam issued from the extended neck of the flask the boiling was stopped and the flask was allowed to cool. Under this treatment the contents of the flask remained unchanged although, in this case, the solution was directly connected through the curved neck with the surrounding atmosphere. This was due to the fact that all particles of dust, including the germs of the micro-organisms, were retained on the curved surfaces of the S-shaped tube. If the neck was cut off the liquid was soon colonised by micro-organisms. In this experiment the air was submitted to absolutely no treatment and nevertheless decomposition of the liquid did not occur, simply because the organisms floating in the air were denied access to it.

Further investigations by Pasteur showed that the content of viable micro-organisms in the air was far from constant and changed according to conditions such as season and place. The largest number of germs is present in the air of towns and inhabited places. The air of fields and forests is less rich in micro-organisms, and finally in the mountains, especially at great heights, the number of these minute living creatures floating in the air is quite insignificant. One may therefore open flasks containing sterile liquids without their necessarily being exposed to infection. In many cases such flasks remained sterile after resealing, although untreated mountain air had been admitted to them.

Pasteur also demonstrated that the air is far from being the only source of infection of organic liquids. The germs of micro-organisms are present on the surfaces of all the objects which we use in the course of an experiment. Therefore all these objects must be meticulously disinfected. Pasteur showed that the appearance of micro-organisms in the experiments of earlier investigators was always due to the fact that they had not carefully eliminated all sources of infection. Thus, for example, Pasteur showed by direct experiments that the source of infection of Gay-Lussac's grape juice was micro-organisms present on the surface of the mercury. In other cases the organisms were derived from incompletely sterilised utensils. If all sources of error are avoided then, as Pasteur demonstrated brilliantly in numerous experiments, infection will be absent in a hundred per cent of cases. Pasteur also succeeded in showing that it is possible to keep even such easily decomposed liquids as urine and blood for an indefinite time without submitting them to heat or any other treatment. It is only necessary to withdraw them from the body of the animal, where they do not contain bacteria, while taking precautions against contamination with germs from outside. Under these circumstances such liquids do not putrefy and may be conserved indefinitely.

Pasteur did not merely aim at getting accurate and uniform results but also at explaining the contradictory data of other authors. He rejected the suggestion that decaying infusions give rise to microbes and showed that, on the contrary, the decay of these liquids itself takes place as a result of the vital activities of micro-organisms which have entered from outside. All attempts to refute this hypothesis and to find a case of spontaneous generation of any particular organism were in vain. From our present point of view this is quite understandable, in that micro-organisms are not simple lumps of organic material as was believed until the time of Pasteur. A detailed study of these very simple living things has shown that they have a very delicate and complicated organisation. It is quite impossible to suppose that complicated structures of this sort could emerge in the course of a short time before our eyes out of structureless solutions of organic substances. This hypothesis is, in essence, just as absurd as the hypothesis that frogs arise from the dews of May or lions from the stones of the desert.

Pasteur's investigations quite understandably attracted tremendous attention among his contemporaries. The complete revolution in biology brought about by Pasteur may be compared with that achieved by Copernicus in astronomy. For, in the one case as in the other, prejudices which had held sway over the minds of men for thousands of years were swept away.

As we have seen above, many generations of scientists and philosophers considered the possibility of spontaneous generation to be an incontrovertible and self-evident truth. The obdurate struggles between idealism and materialism were only concerned with the theoretical explanation of the 'phenomenon'. And now it was suddenly discovered that the 'phenomenon' itself, the very 'fact' of spontaneous generation, was illusory and was based on false interpretations of observations and incorrect conduct of experiments. At the end of the last century and the beginning of the present one the two warring philosophical camps redeployed their forces in the light of this discovery.

Vitalism, the idealistic tendency in biology, had already achieved its most exuberant development by the middle of the eighteenth century. At that time our knowledge of life was so limited that it seemed quite impossible to explain physiological and formative processes without recourse to the activity of some special, mysterious 'life force'. However, at the end of the eighteenth century there was a tremendous surge of great discoveries in physics and chemistry, and from that time onwards vitalism suffered one defeat after another. Even by the second quarter of the nineteenth century it had really almost played itself out. The evolutionary theory of Darwin dealt a final crushing blow to vitalism. It showed the way to a scientific, materialistic solution of the problem of the adaptation of form to purpose in the organic world. After this the concept of a 'life force' became quite unnecessary, it explained nothing and was a purely mystical and meaningless word.

However, the end of last century witnessed a resurgence of vitalism, which now chose the problem of the origin of life as one of its main rallying points. In 1894 I. Borodin wrote "Has not the progress of science in the course of centuries furnished the vitalists to some extent with weapons? Yes, they certainly have such weapons, they hold a trump card in their hand." Borodin meant by this 'trump' the unsuccessful attempts to discover the phenomenon of spontaneous generation. These failures, in his opinion, indicated the presence of an impenetrable barrier between the animate and the inanimate, the complete autonomy of vital phenomena.

Borodin continued:

> *That old woman, the life force, whom we buried with such triumph, at whom we mocked in every way, was only pretending to be dead and now decides to demand some rights to life, prepares herself to start up in a new form. Our expiring nineteenth century misses fire, it misses fire on the question of the origin of life.*

Thus idealism, which, as we have already seen, argued obstinately throughout its whole history in favour of the existence of spontaneous generation, carried out a complete *volte face* on this question at the beginning of the present century. The triumph of the theory of evolution forced the vitalists to regard the problem of the origin of life as the last refuge of the 'life force'. Darwinism might well give a materialist explanation of the ways in which higher organisms develop from lower ones, but the human mind would never be able to understand how life itself came about, because its essence ('entelechy', the 'life force', the 'cellular spirit', etc.) lay at the limit of the capacity of the intellect. We find this in the writings of most of the neovitalists and other idealistically inclined biologists of our century. Thus H. Driesch wrote of the insolubility of the problem of the origin of this vital principle which he called 'entelechy'. Uexkull drew attention to the necessity for a special transcendental factor (structural plan) for the origin of life. L. Bertalanffy denied the possibility of the self-formation of such a system as, in his opinion, an organism must be. E. Lippmann finishes his book devoted to the problem of the emergence of life with the words: "The limitations of the intellect prevent us from penetrating into the problem of life. We cannot understand its essence which appears to be metaphysical." Thus the idealists try to use the demolition of the theory of spontaneous generation as an occasion for proclaiming the impossibility of solving the question of the origin of life on a materialistic basis.

The leading proponents of materialism rejected this approach to the problem right from its inception in the last years of the nineteenth century. They considered that the fact that microbes do not develop spontaneously in organic solutions and infusions was no argument that life has not a material origin.

One of the first to discuss this problem was F. Engels. He remarked that all investigations so far made in this field had been quite limited in approach, dealing only with the problem of plasmogenesis. Pointing out that spontaneous generation (*generatio aequivoca*) was contrary to the findings of contemporary science, Engels ironically remarked that it would be absurd to hope to compel nature with the help of some stinking water to do in 24 hours that for which thousands of years had been required. Thus Engels emphasised that it was not sudden spontaneous generation but a prolonged evolution of matter which led up to the emergence of life.

However, most scientists of that period still took up a mechanistic position and held that sudden spontaneous generation was not only the simplest, but even the only conceivable explanation of the origin of life. In this connection E. Haeckel wrote "To deny spontaneous generation means to accept a miracle, the divine creation of life. Either life arises spontaneously on the basis of some particular laws, or else it has been produced by supernatural forces." This kind of conviction explains the zeal with which many of the exponents of mechanistic materialism flew in the face of the facts to demonstrate the possibility of spontaneous generation. They saw no other way out. As an example one may mention the violent but ill-founded attacks made by the talented Russian publicist D. Pisarev on the work of Pasteur.

Finally, there was no dearth of experimental effort to show that it was possible for living creatures to come into existence suddenly. However, all these experiments, without exception, were utterly futile. The most serious and interesting were those of Bastian. He showed that micro-organisms developed in boiled infusions of hay even when the flasks containing the infusions were opened on mountain tops or after the air entering them had been brought to a red heat. The investigations of Pasteur were consistent with the factual side of these experiments but Pasteur also showed that spontaneous generation of microbes had not occurred in this case either. The spores of the hay bacillus, which was the organism which grew, can withstand prolonged boiling and still remain viable. If the hay infusion is heated in an autoclave to 120° C or boiled twice it, like other organic liquids, will retain its sterility on the admission of uninfected air. In such cases repeated boiling acts as follows: the first heating destroys all the vegetative forms of the bacteria but the spores remain. After cooling, bacteria develop from the spores but succumb to the second boiling without having succeeded in forming new spores.

The outstanding Russian scientist K. A. Timiryazev, with his usual clarity of scientific exposition, submitted these attempts to demonstrate the possibility of spontaneous generation to devastating criticism. In an address which he delivered at a session of the Society of the Friends of Science in 1894 he spoke as follows:

> *When Bastian created bacteria from an infusion of turnips with rotten cheese in the nineteenth century he was, in this matter, just as much of an empiricist as was van Helmont in the sixteenth century, when he created mice from flour and dirty rags. At least I know of no physical or chemical laws which might lead one to prefer the stinking mixtures of the nineteenth century empiricists to the sluttish mixtures of the sixteenth century one. Attempts to produce spontaneous generation in the nineteenth century are not necessarily superior to such attempts made in the sixteenth century; in fact, they are equally far from the basic ideas which characterise the scientific thought of our times.*

Furthermore, while arguing with Borodin, Timiryazev declared:

So you pick out two or three foolhardy adventurers with the ideas and mentality of the sixteenth century, going astray in the middle of the nineteenth century; you see in them the representatives of contemporary science and hail their failure as the misfiring of the nineteenth century'. Is that quite fair?

This impassioned reply by Timiryazev is also fully applicable to the empiricists of the present day, the adherents of spontaneous generation who, according to their way of thinking, are rushing to the defence of materialism and who only delude themselves and others with their experiments. Having been concerned with the problem of the origin of life for many years, I have received and still receive a large number of letters with descriptions of different instances of spontaneous generation which is said to have occurred in the experiments of one or another of the writers. However, none of these experiments need be taken seriously. They are amateurish and the sources of error can easily be established.

From the works on spontaneous generation which still appear from time to time in the scientific literature, one may be selected by way of an example because it concerns the scientist F. Elfving, who is well known for his investigations in the field of microbiology. It was published in 1938 in the journal of the Finnish Scientific Society.

Elfving sterilised dried peas by placing them in a solution of corrosive sublimate (3: 1000) for half an hour; he then washed them with sterile water and allowed them to germinate under sterile conditions in Erlenmayer flasks containing a little water. When the peas grew and the sprouts had developed considerably he killed them by keeping the flasks at a temperature of 60° C for one to two hours. Some days after this treatment by heat he noticed that the water in which the dead plants were lying was swarming with bacteria. From this experiment Elfving came to the conclusion that, in the dispute between Needham and Spallanzani, it was Needham who was right. The substance of the peas which had been killed by gentle heating contained a special 'generative power' which gave rise to new living bacteria. It is easy to detect Elfving's mistake. As was shown by investigations on the production of sterile cultures of higher plants, particularly the experiments of G. Petrov, one can never successfully sterilise seeds by keeping them for this or that time in a solution of corrosive sublimate. This is better achieved by the action of a solution of bromine. There can be no doubt that completely viable germs remained on the surfaces of Elfving's peas. Elfving himself remarked that on the peas "there grew mycelia which were obviously derived from some spore which had survived the treatment with corrosive sublimate". On repeating Elfving's experiments, using bromine instead of corrosive sublimate to sterilise the peas, we were easily able to convince ourselves that under these conditions, as was only to be expected, no development of microbes occurred.

We even find an attempt to rehabilitate Pouchet's experiments and thus to resuscitate the theory of spontaneous generation in the much publicised book of O. Lepeshinskaya, The development of cells from living matter. However, no such attempts have withstood criticism by experiment and, as Terekhovskil pointed out long ago, they are foredoomed to failure. The organisation of any of the living creatures known to us, even the simplest ones, exhibits not only a very complicated structure in the protoplasm, a particular arrangement in space of those molecular complexes which constitute the protoplasm, but also organisation in time, a particular series of biochemical processes which, together, constitute the metabolism. We now know very well that even relatively slight interference can produce farreaching changes in such a system. On damaging protoplasm mechanically or by heat the balance of the metabolism is disturbed irreversibly. This disturbance upsets the harmonious interaction of the synthetic processes and markedly intensifies the reactions of breakdown which proceed in a disorderly way.

It is interesting to note that the hypothesis of spontaneous generation was always applied to those organisms which had only been studied imperfectly at each stage of the development of science. Before Redi's experiments it was applied to various kinds of worms and parasites. It was the same with bacteria before the time of Pasteur. Finally, in our own times, an attempt has been made to resurrect the theory of spontaneous generation with reference to organisms discovered during this period but still poorly understood, the ultramicrobes and filterable viruses. However, this attempt has been a complete fiasco too.

Summing up all that has been said in this chapter, one must emphasise that the very idea of spontaneous generation has been based on faulty observations, accepted uncritically, of the sudden appearance of living creatures in nature or in the laboratory. The possibility of spontaneous generation was assumed by philosophers of every school and persuasion throughout the course of many centuries. They only quarrelled about the theoretical interpretation of the 'phenomenon'. However, as the methods of scientific investigation of living nature became more and more precise, spontaneous generation was gradually relegated to simpler and simpler organisms. Finally the sudden appearance of even the most primitive organisms from inanimate material was shown to be impossible. Thus, to-day, the theory of spontaneous generation has no more than a historical interest and cannot serve as an approach to the problem with which we are concerned.

2

Organic Evolution

Introduction

The naturalists had traveled widely and observed minutely the variation that clearly existed within each species. Members of species, they observed, are not identical, but show variation in size, strength, health, fertility, longevity, behavior, and countless other characteristics. Darwin in particular realized that natural variation was used by humankind in the selective breeding of plants and animals, for, by selection, farmers and gardeners, dog owners and pigeon fanciers, would breed only from the particular individuals possessing the qualities desired by the breeder.

In due course, the key to how a similar kind of selection operated in nature to transform wild species of organisms came to both men, and it arose from the same source. The first edition of a book entitled *An Essay on the Principle of Population*, written by an English clergyman, T. R. Malthus, appeared as early as 1798. The author showed how the reproductive potential of humankind was far in excess of the natural resources available to nourish an expanding population. He showed that in practice the size of populations was limited by lethal factors such as disease, famine, and war, and that such factors alone appeared to check what would otherwise be an expanding population.

Both Darwin and Wallace read Malthus' essay independently (probably the second edition of 1803, often reprinted and improved) and, remarkably enough, recorded in their diaries how they realized that therein lay the key to understanding the cause of the evolutionary process. It was clear to them that what Malthus had observed among human populations was true for populations of plants and animals: their reproductive potential was vastly in excess of that necessary to maintain a constant population size. They realized that the individuals which in fact survived must be in some way better equipped to live in their environment than those which did not survive. Thus it followed that in a natural interbreeding population any variation that increased the organism's ability to leave fertile offspring would most likely be preserved, while the variations that decreased that ability would most likely be eliminated.

The theory Darwin and Wallace formulated on that basis (at first, independently of each other) may be stated as four propositions and three deductions. Both propositions (P) and deductions (D) have since been well corroborated by careful observation.

P.1. Organisms produce a far greater number of reproductive cells and, indeed, young individuals than ever give rise to mature individuals.

P.2. The number of individuals in populations and species remains more or less constant over long periods of time.

D.1. *Therefore, there must be a high rate of mortality both among reproductive cells and among immature individuals.*

P.3. The individuals in a population are not identical but show variation in all characteristics, and the individuals that survive by reason of the particular sets of characteristics will become the parents of the next generation.

D.2. *Therefore, the characteristics of those surviving organisms will in some way* have made them better adapted to survive in the conditions of their *environment.*

P4. Offspring resemble parents closely but not exactly.

D.3. *Therefore, subsequent generations will maintain and improve on the degree of adaptation realized, by gradual changes in every generation.*

As a result of the weight of evidence presented by Darwin in his famous book of 1859, *On the Origin of Species by Means of Natural Selection, or the Preservation of Favoured Races in the Struggle for Life,* biologists became convinced of the value and truth of the theory of evolution that he and Wallace proposed. Since then, scientists have closely investigated the processes involved in the different propositions and deductions, and geneticists have come to understand the mechanism that accounts for the origin of variation and the transmission of characteristics. Of direct interest to students of human evolution are the concepts involved in deductions 2 and 3, for, while human genetic processes are no different in kind from those of the rest of the animal kingdom, the selective factors that caused human evolution were unique.

Since Darwin and Wallace set forward their theory of the operation of natural selection, enormous advances have been made in many fields of biology that affect our understanding of the evolutionary process. Most important among these is the science of genetics, which has provided the necessary underpinning for Darwin's ideas. Since the 1930s advances in morphology, paleontology, systematics, cytology, genetics, and molecular research have been combined to form a basis for understanding the process of evolution, which has been called the *Synthetic Theory*. In spite of all this additional knowledge, it is believed that the directing force in evolution is still Darwinian natural selection, and the ideas Darwin developed have not so much been superseded, but rather they have been confirmed, broadened, and given strong and well-tested foundations.

Since the evolutionary process underpins all life forms, animal and plant, it also underpins human life and human knowledge. Because we are a product of evolution, all human activity and understanding will find its roots in this extraordinary process. E. O.Wilson argues persuasively that the fact of our evolutionary origin can give a unity to all human knowledge an underlying theory of knowledge and belief. The importance to us all of Darwin's astonishing insight cannot be overstated. In the coming century it will transform our lives.

Natural Selection and Fitness

It is clear that only a proportion of individuals in a population survive long enough to reach maturity and, in their turn, bear offspring. The environment itself determines the fate of each and, in destroying a proportion, selects the remainder. Through its effect upon each individual the environment controls to a decisive extent the direction and rate of evolution, and for that reason it may be considered to be one creative factor in the process of evolutionary change.

Although natural selection acts on individuals, it is the population that evolves, since the genetic plan of an individual remains constant throughout its life. A novel genetic plan arises only as a result of sexual reproduction. Not only are successive generations therefore necessary for the introduction of new gene combinations, which constitute another creative factor, but they are in fact the source of variation on which natural selection acts.

A series of successive generations reproducing sexually relates individuals not only through the dimension of time but also in the dimension of space. Animals must find a mate among their contemporaries, and if they mate more than once in a lifetime (as mammals do), sexual relationships will be widely spread. Thus, the unit of evolution, the breeding population, includes all the individuals able to mate with each other. The size of the population may vary, but it is the breeding unit with its network of sexual relations that evolves in the course of time.

The *fitness* of such a population, in the biological sense, requires not only the ability to cope with the existing environment and to reproduce, but also the potential to evolve in the future in response to environmental change. This potential requires not only genetic stability, which reflects the broad stability of the environment, but also genetic variability (consequent upon sexual reproduction), which reflects the instability of the environment. That is to say, a population cannot afford to vary greatly in a stable, competitive, and hostile environment, for random variation may be lethal; the population must remain well adapted. At the same time, the population must be able to evolve in adaptation to environmental change. This necessary genetic stability, accompanied by flexibility in the form of adaptability, is the basis of Darwinian fitness, and the balance struck between these two factors determines how fit a population is and the nature of its adaptation.

The dynamic stability of all the genetic components of a population (called the *gene pool*) makes possible adaptation to the environment as well as the modification of such adaptation in the presence of environmental change in the future. The absence of such modification can result in extinction; a proper balance must be found between stability and flexibility, and it is an alteration in the form of this balance that, among other things, characterizes human evolution.

It is clear that every characteristic, every complex of characteristics of the individual, its anatomy, physiology, and psychology, contributes to the biological fitness of the population, and it is in this sense and this sense only that a particular characteristic is of evolutionary significance. It is the contribution characteristics make to the population's fitness that results in their selection, in their survival. It is the population that evolves, not the individual.

At the same time, since all parts of an organism require energy for their maintenance, any part that ceases to have a useful function will be lost in the process of evolution. Not only any part but any process will also be lost. For example, color vision is believed to have evolved among reptiles and then been lost in the very early period of mammalian evolution, probably as an adaptation to nocturnal life; it appeared a second time in the evolution of the primates, but most other mammals cannot see the full range of color because it has not been selected again during their evolution. Thus, we do not often find characteristics without functions, a fact that may well apply to so-called vestigial characteristics; it seems probable that they have at least a reduced function, or are in the process of being lost.

The function of a characteristic can therefore be understood fully only as a process that contributes to the overall reproductive advantage of the population in which it has evolved. The function of any characteristic that cannot be interpreted from this perspective cannot be said to be properly understood. It follows that in order to understand human evolution it is desirable to consider the function of each new characteristic that evolved and to discover how it bestowed upon the population in which it became established a greater probability of survival in a changing environment.

How populations have survived by changing their nature is the story of evolution. The concept of fitness involves both adaptedness and adaptability, but, like evolution itself, though it may be elucidated in the past, it can only be surmised in the present.

We have said that natural selection operates only at the level of the individual, but it has been claimed by Darwin and others that under certain conditions it will also operate on social groups. This may possibly be the case among some higher primates and especially human groups, where social bonding and social interactions are so much closer and more vital than in any other species. This view is presented by Grant, who suggests that if troops or bands of early humans came into competition for a foraging territory or some other essential resource, selection operated on the band as a group rather than on its individual members. The success of a band in competitive encounters is determined by, among other factors, the intelligence of its leaders and members, its technology, communications, and numerical strength. Grant suggests that group selection could come into play when competing groups differ genetically with respect to these characteristics. This might well be so, especially if the band itself is small and is totally destroyed either directly in warfare or indirectly through starvation. This was probably not a very common occurrence, but almost certainly a small degree of group selection could have operated in human evolution, which would have given significant advantage to groups with a high level of intelligence, good coordination and communication. Wilson, whose theories of the biology of social groups make the concept of group selection redundant, nevertheless accepts the possibility of group selection as exclusive to human groups, particularly where warfare is practiced with a degree of genocide.

In his book of 1871, Darwin also recognized another kind of selection which he described as a special kind of natural selection and which he named *Sexual Selection*. He described two kinds of sexual selection which operated to produce new anatomical characters. These kinds of selection were the product of (1) rivalry between males for females, and, (2) the choice of mates made by the opposite sex.

The characteristics produced were not adaptations to the external environment but adaptations to the social environment within a population; interactions related to reproductive activity.

Genotype, Phenotype and the Environment

The *genotype* of an individual organism is its heredity the factors that determine the path development may follow in different environments. The determinants are called *genes*, and they take the form of a complex chemical substance known as *DNA* (*deoxyribonucleic acid*). This substance, which forms immensely long spiral chainlike molecules, is found in the nucleus (the controlling center) of every living cell. The DNA of the male and female gametes (the sperm and egg), together with smaller bodies containing DNA outside the egg nucleus called mitochondria, when combined in the fertilized egg cell or zygote, determine the form and structure of the new individual into which the zygote will develop. From that time onward, as the zygote divides many times, the DNA molecules are copied exactly, and similar copies lie within every cell of the growing individual's body. Once the genotype is determined at fertilization, therefore, it is fixed throughout the life of the individual organism. It is a highly condensed coded bank of information, which during the development of the zygote determines the kind of organism that will grow.

We can with some certainty postulate that there are two main categories of active genes within the genotype of each individual. The first, called *structural genes*, determines the structural ground plan of the organism. The second, called *regulatory genes*, determines the rate of growth of the different parts of the body, the different organs, including the time of the onset of sexual maturity, which signals the end of physical growth. It seems clear that alterations in a very few regulatory genes (affecting, for example, the size of certain structures or organs) can have quite a marked effect on adult structure without fatally altering the functional integrity of the organism.

The *phenotype* is the whole individual, the manifest characters of an organism, the discrete biological unit, the human being or the ape. It is formed as a result of the living interaction of the genotype and the environment, an interaction called assimilation or growth. The phenotype is neither constant in form and structure nor permanent. Through its inter- action with the environment, the genotype determines not only the phenotype's mature form (its *morphology*) but also its total growth pattern: that is, its form at all ages, together with its growth rate, its capacities, and its lifespan. While the genotype is constantly copied in cell division, its form does not change, but the phenotype itself does change as it grows from a fertilized egg into old age. When the phenotype dies, as we all do, the fit genotype will survive into future generations-immortal.

Thus the phenotype is a temporary home created by the genes; as Dawkins has put it, phenotypes are "survival machines" created by and for genes, and they are mortal and dispensable.

Variation in the genotype occurs between generations as a result of sexual reproduction. In the process of fertilization, a single set of paternal genes is combined with a single set of maternal genes to produce a novel genotype. With the exception of identical twins, no two individuals of *Homo sapiens* are ever likely to be similar in spite of their vast numbers, so efficient is the shuffling of the pack of hundreds of thousands of genes that occurs in sexual reproduction. By combining different genotypes in fertilization, sexual reproduction is an essential source of genetic variation in evolution, and the process itself is believed to have been selected as a reproductive mechanism for that reason. It is through selective breeding that plant breeders can create such remarkable phenotypes as the hybrid roses we enjoy in our gardens. Originally created by sexual reproduction and selection, they are, in turn, preserved by asexual vegetative propagation in which no further significant variation of the genotype occurs.

However, although sexual shuffling is of great importance, it is not the fundamental source of genetic variation. There is also the phenomenon of so-called "spontaneous" change in the genotype (*mutation*), which results from "imperfections" in its natural and ordered reduplication during cell division. This valuable and inherent "flaw" in genetic replication is most common in the production of the paternal and maternal gametes, when the paired gene sets of the normal body cells are separated. In this process of division (*meiosis*), the DNA chains undergo stress, which very often results in chemical modifications. This kind of mutation is random and therefore usually lethal, for no highly complex mechanism is likely to benefit from random interference.

Thus, the mutations that allow survival of the gametes (let alone of the individual) and that are subject to natural selection, are themselves already selected by the need for internal coordination in the gamete and in the zygote. This internal selective process limits the kinds of mutations available to be selected by the environment. The nature of organic mechanisms is such, however, that slight changes that might improve adaptation in a changing environment are possible. Organic characteristics (for example, growth rate, body temperature, size, and so forth) vary continuously (not discontinuously, in jumps), so small adjustments in the mean or average form of these characteristics may be advantageous at any time. It follows that spontaneous mutation, if it has very slight effect on the phenotype, can be of value to an individual, and if it is, it may spread by the aid of sexual reproduction to a whole population quite rapidly. But the value of such mutations can be realized for the population only by natural selection, the selective interference of the environment upon the gene pool.In practice, the environment can interfere with the genotype in two ways, which blend into each other:

1. Random interference from the environment is perhaps the least common. The most obvious example is the radiation that enters the earth's atmosphere from outer space and penetrates organic substance. It may be *mutagenic*; that is, it may cause mutations in the genotype of the cell that it penetrates. Should this cell be a gamete or newly fertilized egg cell, the radiation may either kill the cell or, by changing the chemical structure of one or more individual genes that do not lethally affect cell function, cause an increase in variation in the gene pool. Atomic radiation is powerfully active in this way, and although artificially created and localized, its effects on the genotype are random. Such influences are not likely to be a significant source of variation in organic evolution over the immense periods of geological time.

2. Most of the other ways that the environment acts upon individuals are more or less selective, and they may be either lethal or sublethal. Lethal interference of a particular type (for example, famine) may kill some kinds of individuals (for example, thin ones) more readily than others, and for this reason it will change the nature of the gene pool of the population (individuals with reserves of fat are selected). Sublethal interference will also affect the nature of the gene pool if it affects the reproductive capacity of any individuals as a result of a particular characteristic they possess (for example, disease or displacement of the uterus may result in abortion). Such selective interference is enough to result in evolutionary change, even if all the individuals survive; it is only necessary for them to vary in the rate at which they reproduce themselves.

The environment strongly affects the growth and development of individual phenotypes because they show a certain amount of developmental plasticity or *adaptability*. This plasticity is itself adaptive and as such helps to ensure a high rate of reproduction in the face of environmental variation. The effect of the environment upon the gene pool will ultimately depend on the intensity of the interference, or *selection pressure*, as it is usually called, and the degree of adaptation evolved by the species in response to this pressure.

It should be noted that the individual (and, of course, the population) will alter the environment to some small extent merely by its existence. While the individual genotype cannot change in the face of selection pressure, which may take the form of over-population, the gene pool can and does. Such change is evolution; failure to change may result in extinction of the population.

It must be recognized that as a correlate of the theory of evolution by natural selection an element of competition is invariably associated with organic life. Because organisms reproduce at a rate much higher than that necessary to maintain a stable population, there is always competition within a single population for the available food and space. In this way, every individual changes the environment. Similarly, there is competition among populations of the same species, as well as between species. It is noteworthy, however, that members of a single species are in more immediate competition because they require the same food and habitat, and therefore they alter their own environment more than members of different species that are adapted to different parts of the environment or ecological niches. It should be added, however, that in advanced social species of the higher primates, cooperation is superimposed upon competition within social groups.

Natural selection resulting from competition within species (*intraspecific* competition) is therefore as much a cause of evolutionary change as selection resulting from competition between species (*interspecific*). Intraspecific competition is a permanent feature of organic life, as is natural selection, and a noncompetitive situation cannot exist for more than an extremely short time because the reproductive potential of species is always greater than that which is required to maintain even an expanding population in times of beneficial environmental change. This intraspecific competition means in effect that the evolution of further adaptation may occur even when the external environment is fairly constant. Thus, evolution within one population will, of course, come to change the environment of its neighbors.

It is clear that the total interaction among individuals, populations, and their environment is intricate, and the totality of living plants and animals upon the earth's surface forms a dynamic system of great complexity. This complexity is seen to be immense when we consider further that among animals the connecting link between the environment and the phenotype is the medium of behavior, since what an animal *does*, as well as what it *is*, determines its ability to survive.

Mutations Small and Great

The term mutation, briefly introduced in the previous section, leads us to consider the sources of such variations in more detail. There are two main categories of mutation or changes in the nuclear DNA which we shall briefly summarize.

1. Genetic or DNA Mutations. These cannot be observed but can be inferred by a number of different means. They are responsible for phenotypic change (both structural and physiological changes) and are the common and basic origin of variation which powers the process of evolution. Though these mutations may in themselves be very small alterations in DNA sequencing, they can ultimately generate both small gradual evolutionary change and the relatively fast and highly significant developments which are so striking in the fossil record.

2. Chromosome Mutations. These are due to observable rearrangements of chromosomes, their duplication or their loss. Because chromosomes pair and segregate during meiosis any major change in chromosome structure (if present in only one member of a chromosome pair) can effectively form a breeding barrier between otherwise related individuals or populations. Indeed, the lack of match between the chromosomes of different species is one important barrier between them, should mating take place. During meiosis, sections of chromosomes may also be reversed or moved to a different part of the structure. Chromosomes may also join up together, or split to form a new and separate pair. The type and number of chromosomes which typify a species are called its *karyotype*.

Chromosome mutation is of interest to us because one of the striking differences between humans and the great apes is that the great apes have 24 pairs of chromosomes and humans have only 23 pairs. Apparently two of the smaller rod-shaped chimpanzee chromosomes have fused to form our chromosome No. 2 at some point along the human lineage since the last common (i.e., human-chimpanzee) ancestral population existed. The important question then is to consider what effect this major mutation might have had. A number of suggestions have been put forward.

In the first place, the change may have had almost no effect on the phenotype, because the original genetic material is still present and intact in the nucleus of each cell. However, genes do act differently according to their position in the chromosome (the *position effect*) and it does seem likely that this development must have had some effect on the phenotype during human evolution. Some scholars feel that this change may have had a very significant effect on the course of evolution. It has been suggested that it may have been connected with the changes in the shape of the pelvis which resulted in efficient bipedal walking. Others have suggested that it was a late development and brought about the changes which we associate with modern humans, and which distinguish us from earlier forms of *Homo*. What is clear is that it must have been beneficial, since it spread among the entire human population.

There is another mystery. If the change occurred first in one individual, that individual might have been sterile unless it found a sexual partner with the same mutation. It has been suggested that this could only have happened in a very small community where inbreeding was common. Such an event would be rare and to be selected it would need to carry a significant selective advantage.

Environmental Change and Evolution Rate

Environmental change is the prime determinant of alterations in the gene pool of organic populations. Environmental change may widely vary in form, but its ultimate cause is change in the earth's climate. A long-term climatic change in a geographical region directly affects every single living organism within that region. In addition, every organism will be indirectly affected by the change in every other organism that the change in climate causes. For example, a steady lowering of rainfall in a particular region will change the vegetation from plants adapted to a relatively high rainfall to plants adapted to a relatively low rainfall. Such a change will immediately affect all herbivorous animals, which may be unable to find enough of their usual food plants. If they survive, such animals will have slowly adapted their behavior and digestive processes to the new food plants that become dominant. In turn, their carnivorous predators will also be affected. Change in rainfall may also, for example, affect the transmission and occurrence of parasites, for they may depend on damp conditions to survive the period of transmission between their host animals or plants. Drinking water will perhaps become scarce, and animals may survive only by an adaptation in their water-storage capacity so that they require only one visit to a water hole per day. Ultimately, the climate may become subject to occasional extremes of drought, which may be responsible for the extinction of certain species herbivores, carnivores, or their parasites.It is clear that a single change in one component of a community of plants and animals may have very complex and far-reaching effects on the whole nature of the environment. Adaptive adjustment among all the different populations of organisms may be a lengthy process, for every change in one species will affect the environment and in turn effect change in every other. In stating, therefore, that the ultimate cause of environmental change is a change in climate, we refer to the whole causal chain of adjustments that result from a change in the prime determinants of the earth's environment: the level of solar radiation, the nature of the earth's crust and atmosphere, and the earth's rotation.Evolution, which is the resultant of genetic variation and environmental change, can therefore occur in the following ways:

1. In a more or less constant environment, a population may evolve a better adaptation to it. Such evolution may, however, be limited in extent.
2. From such a more or less constant environment, a population (usually part of an existing population) may expand into a neighboring and different environment to which it may prove to be already adapted to some extent. New adaptations may then follow.
3. The environment may change, and the population may adapt to the change. That is of course the most common situation; it results in situation 1 and very often causes the changes in distribution that result in situation 2. It is, however, worth noting that environmental change may not

necessarily result in evolution, since, if the change takes the form of movement of climatic belts on the earth's surface, the fauna may move with the belts, so no new adaptation is necessary. Such a situation would be particularly likely in animals like monkeys or the herbivores of the plains, which are highly mobile and would move within shifting climatic belts without any modification in their way of life.

It has been claimed that a change in gene frequency resulting not from selection by the environment but from random genetic variation in small populations temporarily relieved of selection pressures, known as *genetic drift*, can result in significant evolutionary change. Evidence that this phenomenon of drift has been a significant factor in primate or human evolution is not at present impressive. It may, however, have played an important part in the evolution of other groups of plants and animals. The matter does not require further discussion here. This does not mean, however, that we understand the selective factors that have caused the evolution of every characteristic of living species.

Although Darwin did not have the knowledge of genetics that we have today, he did visualize the process of evolutionary change to be gradual and the product of a succession of small changes in the germ-plasm being subjected to natural selection. It was not until the turn of the century that the Dutch botanist de Vries drew attention to the importance of major mutations in the evolution of new species. In his studies of the evening primrose, *Oenothera*, he observed such mutations and recorded the appearance (within a few generations) of many entirely novel varieties and species, which bred true.

De Vries believed that such mutations must be the basis of evolutionary change, and that natural selection operating on minor variations could never be sufficient to bring about the varied evolutionary radiations of plants and animals that today cover the earth.

Since then our knowledge of the evolutionary process has grown enormously. The modern version of Darwin's original ideas, the *Synthetic Theory*, which incorporates modern genetics, recognizes both gradualism in evolutionary change and the possibility of episodes of very rapid evolution. The hypothesis that small isolated populations, in particular, may undergo rapid change has been part of the synthetic theory for 50 years.

This is important because a lively controversy exists today between those who believe that the synthetic theory still represents an adequate, reasoned explanation for the process of evolution, and those who believe that changes are needed in it and that a much greater weight must be given to the phenomenon of rapid change interspersed by periods of adaptive stability (stasis). This concept of *punctuated equilibrium* has been developed as constituting a central process in organic evolution by some contemporary scholars. They believe that steady, gradual change is a rare phenomenon, and imply that the synthetic theory as it stands cannot account for very rapid change.

It is the view of this writer, and still the consensus among biologists, that there is no need to modify the synthetic theory or to postulate major and highly improbable adaptive mutations to account for periods of rapid change. Simpson discussed what he termed quantum evolution in a number of books. He described the means by which natural selection can bring about a very rapid phase of evolution between two adaptive zones that are more or less discontinuous. The evidence for such events is well manifested in the evolution of many groups. Such a quantum change may have occurred at a number of points in human evolution, but most probably between the earlier adaptation to the forest woodland zone and the later adaptation to the savanna. It is clear, however, that the synthetic theory as generally presented does allow for apparent discontinuities in the fossil record, which represent periods of rapid evolutionary change; and that this rapid change can be understood without modification of the synthetic theory. We can conclude that periods of gradual change and periods of rapid change may have often occurred and that the latter may have played an important part in the evolutionary process. We may find that a lineage passes from a period of almost complete stasis through gradual change or alternatively a lineage may take a quantum leap. Any sequence of these different rates may occur. A very detailed study of a complete and very well-dated fossil record would, however, be required to identify each part of the process.

The rate of evolution is ultimately related to its two causative factors: genetic variation and environmental change. Environmental change is the prime determinant of the evolution rate, but in times of fast change the maximum rate will depend upon the amount of genetic variation available from mutation and from the sexual shuffling of genes. How is the mutation rate itself determined? It varies considerably among different animals and plants, and it seems most probable that the rate is selected in evolution like any other characteristic. A fast rate will allow fast evolution, yet will produce a higher ratio of ill-adapted members in a stable population. On the other hand, in times of rapid environmental change a slow rate might block sufficiently fast adaptation to permit survival. Like every other characteristic, the mutation rate of a population is adapted to the environment, and in particular to its overall rate of change. Genetic stability and variability are important features of biological fitness.

Variability, Speciation and Taxonomy

The sources of variation operating in animal populations and described above result in a great variety of forms in any particular population. The fact that every individual has a slightly different genotype (unless a monozygotic twin) and has been subject to slightly different environmental stress means that every individual, when mature, will be different in a large number of features. Beyond this we can expect variability due to age: babies and mature adults vary not only in size but in bodily proportion, and in the presence and absence of certain features. The two sexes also may vary considerably, not only in the generative organs but in other features as well the secondary sexual characteristics, which may affect stature, weight, and even shape to a surprising extent. For example, male gorillas are almost twice as large as females, and the proportions of parts of the skull differ between them. In the past, paleontologists have made the understandable mistake of classifying rare fossils of different age and sex as different species or even genera.

Beyond the variation that we can expect within a population, there will also be variation between different populations of a single species. Populations living in different geographical areas will experience slightly different environments, for no two places on the earth's surface are exactly the same. These different *geographical races*, as they are called, vary in adaptation to the unique features of their differing environments: this is geographical or racial variation. Local races at the opposite extremes of the range of a species may well show very considerable morphological differences. In this case we can only be sure that they belong to a single species by tracing the continuity of populations over the area of land between them. If a single species is represented, then the continuity of individuals will be associated with a continuity of morphology, which implies that genes are being exchanged between neighboring populations. This exchange of genes between populations is termed *gene flow*, and it maintains the integrity of the species.

Should some barrier become interposed between parts of the species range, then gene flow will be interrupted, and different populations will become geographically isolated. This *geographical isolation* will allow the separated populations to vary independently, perhaps in different ways.

If the isolation is complete and remains so for a long time, the independent adaptations of the two groups of populations may well result in morphological divergence to a point where they are very different. If the difference is so extensive that on coming into contact again at a later date no further gene flow occurs, then the populations will have *speciated*: that is, one species will have split into two. The necessary conditions for speciation are the variability that is present in all species, the operation of natural selection, which is equally a feature of all organic life, and geographical isolation. This last feature is the immediate cause of speciation.

The appearance of geographical barriers is not an everyday occurrence. Changes in climate, and particularly rainfall, may result in the increased size of rivers, lakes, and seas to form impenetrable barriers to certain animals. Tectonic movements may result in the merging or separation of continents, in new mountain ranges or deep rift valleys, with associated changes in climate. Far less extreme changes in climate and environment may result in what amount, in effect, to geographical barriers, depending on the mobility and behavioral plasticity of the species. But the formation of the barrier is the reason one species splits into two, (called *allopatric speciation*), which is the way a single evolutionary lineage also splits into two, and so may bring about a whole new evolutionary radiation.

This discussion leads us to the definition of species as *groups of interbreeding natural populations that are reproductively isolated from other such groups*. The critical feature of successful speciation is failure in interbreeding the absence of gene flow and the buildup of morphological divergence. Biological species, or *biospecies*, are the groups that fulfill this definition, the species of plants and animals that surround us. The nature and recognition of fossil species, sometimes called *chronospecies*, present some special problems that are discussed below.

We have seen in this section that a biospecies is a natural unit, called a *taxonomic unit*, that can be defined and recognized in nature, generally without much difficulty. Species are often divided into smaller units called *races*. The term race usually refers to a group of populations with a certain number of (usually superficial) features in common that distinguish them from other such groups. However, since they are not isolated by genetic discontinuity, their boundaries are not easily recognized, and the definition of particular races is therefore always a matter of discussion and often of disagreement. Important and large racial groups are sometimes called *subspecies*, while minor ones are called *local races*. Almost without exception, animal and plant species can be divided into these lesser units according to their geographical range and variability. The presence of different races is a typical characteristic of living species and represents their potential for further evolution and speciation.

Finally, species themselves are grouped into units called *genera* (singular *genus*), which are groups of species with major adaptive features in common. Like races, these units are subjective and do not always represent easily defined natural categories.

Two Latin names, or *nomina*, are used to label each species; the first is the generic name, the second the species name. Thus, *Homosapiens* is the name given by Linnaeus in 1758 to mankind: it indicates that we belong to the genus *Homo* and the species *sapiens*.

The conventions of zoological classification require that at each level the taxonomic units, or *taxa* (singular: *taxon*), are grouped into taxa of a higher level. Thus, genera are grouped into families, families into *superfamilies*, superfamilies into *orders* (such as Primates), orders into *classes* (such as Mammalia), classes into *phyla* (such as Chordata), and so on. This system of classification is essential as an informative basis of communication among zoologists, and many of these terms will be used in the following pages.

We must now consider how the taxonomist goes about classifying populations of organisms. This is an important question, since it affects the resulting classification that he builds. The assessment of relationships by the study of comparative anatomy is a complex question. The simplest approach is that of *phenetics*, in which morphological relationships are analyzed by giving all characteristics compared equal weight. The analysis should be based on a large number of characteristics, and the results produced simply group like species together and separate them from those that are less alike. This technique produces a taxonomy which is useful but does not reflect evolutionary history, only present morphology. It is convenient when trying to classify forms such as bacteria, which have no fossil evidence that can help in understanding their actual evolutionary relationships.

Table : The Use of Taxa

	Taxon	
Taxonomic category	**Example 1**	**Example 2**
Kingdom	Animalia	Animalia
Phylum	Chordata	Chordata
Class	Mammalia	Mammalia
Order	Primates	Primates
Suborder	Anthropoidea	Anthropoidea
Superfamily	Hominoidea	Hominoidea
Family	Hominidae	Gorillidae
Genus	*Homo*	*Pan*
Species	*H. sapiens*	*P. gorilla*

The *phylogenetic* approach attempts to assess relationships as they actually exist, just as human genealogists do. Thus, the taxonomy will take into account fossils and their age; it will take into account similarities between groups, those that are shared with ancestral forms and those that are more recent developments; those that are specialized in the lineages being investigated and those that are primitive and widely shared. The phylogenetic taxonomist will take into account all characteristics, but will give greater weight to the specialized, recently evolved ones than to the ancient ones. Thus, he will weigh his characteristics to achieve a classification he hopes will reflect actual evolutionary relationships. For example, the distinctions between the different means of nourishing the embryo are given great weight in the classification of the two great subdivisions of the class Mammalia (marsupials and placentals), but these characteristics are given no weight in the classification of the order primates, or any other order of mammals, as they are widely shared and show no particular specializations in these orders. At this lower level, dentition, which is more variable, is given greater weight. It is important to note that in arriving at these judgments, taxonomists are not looking at characteristics separately, as in a phenetic analysis, but at the total morphological pattern of the population in its functional aspect.

Biologists using this method arrive at a generally acceptable consensus, but the process of classification requires both experience and subjective judgment, so that some scientists have described the process more as an art than a science. A refinement of this approach, in which personal judgment and experience is believed to be reduced, takes the form of a method of analysis called *cladistics*. In this approach the taxonomist clearly divides the characteristics of the specimens he is studying into either ancestral (*plesiomorphic*) or derived (*apomorphic*) characteristics. Thus, all relationships within a particular taxon are determined by a study of individual derived characteristics only, because the ancestral characteristics are necessarily shared and therefore not taxonomically significant. The more derived characteristics that are shared between units of the group under study, the closer their relationship. Evolutionary change over time implies the substitution of ancestral by derived characteristics.

As we have seen, the process of change over time is inconstant; the rate of evolution varies, some characteristics evolve while others remain static, and in a few cases some reversal in the direction of change may occur. For these reasons, the conclusions about relationships reached by cladistic analysis are usually not believed to conform to actual evolutionary history, which is to say, they do not represent phylogenetic relationships, but practitioners claim that the results are the best that can be obtained and that they create the most useful classificatory schemes. However, it should be pointed out that cladistic analysis is not an entirely objective process. Human judgment is required in distinguishing ancestral from derived characteristics and in isolating those characteristics, from the functional complexes in which they exist, which are required for quantification. Thus, the data are used to establish the closeness of morphological relationships within major groups and only later used, if ever, to establish lineages. Some researchers never go beyond the establishment of morphological relationships and altogether avoid making deductions about evolutionary lineages.

Our view here is that cladistic analysis, through its rigid methodology, is a useful indicator of relationships, and it will generate a branching sequence. An evolutionary biologist, however, must take the step of formulating chronological sequences of fossils if he wishes to understand the evolution of a particular group of organisms. We shall be concerned with the overall pattern of the evolution of the hominids rather than with taxonomic procedures and conclusions. It should be pointed out, however, that different views on primate taxonomy can usually be attributed to the different theoretical approaches of the taxonomists who prepare them.

Using a phylogenetic taxonomy, the hierarchy of taxonomic terms indicates to an evolutionary biologist a summary of deductions about the evolutionary history of a group. Cladistic analysis and the resulting taxonomy (which treats all groups as "contemporary" because the chronological data are not considered to be relevant), though it effectively indicates the degree of relationship, cannot reflect our fullest understanding of the evolutionary process.

Fossil species are properly called chronospecies, which implies that they have a dimension of time as well as spatial and morphological variability. They represent a segment of an evolving lineage, a biological species extended through a long period, with the variation due to evolution added to that which we would expect of any population at a given point in time. The main difficulty that arises from the concept of the chronospecies is that *successive chronospecies*: in a single lineage are continuous and not discrete; there is no real natural boundary between them. Those that represent two contemporary lineages can easily be distinguished by the methods usually employed to recognize biospecies, but where they succeed each other in a single lineage, the boundary must be artificially drawn by the taxonomist. The result is that many different opinions exist as to the best place to draw these boundaries; the same fossils are given different names by different authors and their nomenclature becomes confusing.

Homeostasis of the Individual

It is necessary to consider the interaction of the individual and the environment, in order to understand how one is related to the other. An individual organism is a very delicate, complex, and clearly very unstable living system, since it can stand only limited environmental variability; a momentary electric shock, a heavy blow, extremes of temperature or humidity, or shortage of oxygen may destroy the living system entirely. Its survival depends upon its property of self-regulation.

A living organism is a self-regulating dynamic system that maintains a more or less steady state or equilibrium, both within itself and between itself and its environment. Cannon proposed the term *homeostasis* for the self-regulating property of organic systems. The particular character of an organic homeostatic system arises from the fact that it is an open system that depends on an energy supply obtained from the environment by chemical interchange for its continued existence, yet at the same time must maintain itself as an integrated and discrete mechanism separate from the environment. Furthermore, the system is maintained throughout the growth and development of each individual, and during this period the whole cellular structure is being continually broken down and replaced as it matures. The same body in infancy and old age has little in common beyond its genotype and a dynamic system of a particular kind. Human personality, the continuing identifiable nature of an individual, is an aspect of this system.

Cannon has described in his book, *The Wisdom of the Body*, how the homeostatic physiological mechanisms of the human body maintain what has been termed its "internal environment" at a constant level. He has described how the body maintains constant (within narrow limits) its water content, salt concentration, level of sugar, fat, and protein in the blood, oxygen supply, temperature, and many other features of its organization. The efficiency of these systems is remarkable; for example, humans can survive dry-heat temperatures up to 128° C without an increase in body temperature above normal (37° C). Arctic mammals can similarly survive 35° C below zero without a drop in body temperature. This particular homeostatic mechanism is characteristic only of living mammals and birds, and it is of the utmost importance in the story of human evolution. It enables an animal to survive changes in the external environment that otherwise would destroy its delicate chemical systems. Animals with a limited range of homeostatic mechanisms are limited to more or less constant environments; animals with a wide range of homeostatic mechanisms can occupy unstable environments and survive external conditions that in no way approximate their internal environments.

Thus, marine organisms live in a relatively constant environment of water containing salts, food, and oxygen. For deep-sea animals, changes in temperature associated with the seasons are negligible. In the course of the evolution of marine creatures into terrestrial vertebrates, homeostatic mechanisms have been evolved that maintain their internal environment so that it will approximate that of their marine ancestors. Every living cell of the body of mammals is bathed in a fluid called lymph. The composition of this watery matrix is kept constant by the diffusion of salt, proteins, sugars, oxygen, and other substances from the blood vessels. Thus, we can survive dry heat and cold because the living cells of our bodies are preserved in a saline environment of constant temperature and unvarying composition.

The homeostatic mechanisms of mammals are numerous and complex; they may operate in varied ways such as sweating or suntanning (the former lowers the body temperature, the latter lowers the penetration of ultraviolet light). Both these adjustments to hot, sunny weather are methods of coping with environmental change and, as such, increase the chances of survival of their possessors. These, however, are not the only kinds of homeostatic adjustment that is made, for the maintenance of the organism depends on chemical interchange with the environment. Thus, a low water content will cause the sensation of thirst and lead to the consumption of water; a high carbon dioxide content will cause an acceleration in the respiratory rate to increase CO_2 loss and oxygen intake from the atmosphere. Food is required to maintain sugar, fat, and protein levels. The sensations of thirst, breathlessness, and hunger are therefore aspects of the homeostatic mechanisms involved, and so in turn are the behavioral responses necessary to satisfy these needs. If there is not food, water, and oxygen in the environment surrounding the organism, the organism must go to them. Behavior, then, is part of the homeostatic mechanism; it is the process whereby animals satisfy their need to maintain their internal equilibria. Indeed, the maintenance of this steady state is survival, and animal behavior exists simply to that end and that of reproduction.

An extension of the evolution of such behavior is the use of *artifacts* to increase a species' adaptive potential. Animals that build nests or dig burrows are artificially modifying the immediate environment to make their survival more probable by avoiding extremes of heat, cold, or drought, or by gaining protection from predators. Human material culture, particularly in the development of *facilities*, functions precisely to control body heat and energy flow. Fires permit adaptation to cold climates, and space suits maintain atmospheric pressure and so make it possible to survive in a vacuum. Thus, one of the most significant features of human evolution is the development of cultural extensions of homeostatic mechanisms that allow adaptation to environmental extremes that a cultureless primate could not survive.

Physiological and behavioral homeostatic mechanisms are not the only means by which an individual maintains a steady state in relationship with the environment. Another kind of homeostatic adjustment is discernible during individual growth and development; it is called, appropriately, *developmental homeostasis*, and is a mechanism, as yet not fully understood, whereby the development of the phenotype seems to involve some self-regulating capability that maintains its integrity in the face of environmental variation. Like any self-regulating mechanism, developmental homeostasis is based on a factor of stability (due to the genotype) and a factor of variability, which is the plasticity of the phenotype during its development. Examples of such *plasticity* are found among lower animals and plants more easily than among vertebrates, which have other methods of dealing with environmental change (that is, versatility of behavior). Yet the plasticity of individual vertebrate development is clear. For example, our physique is affected by the kind of life we lead; muscles enlarge with use or atrophy as a result of disuse; fat deposition depends upon food, activity, and climate as well as upon genes. The digestive processes can become adjusted to different diets, and jaw strength and size respond to the demands made upon the masticatory apparatus by the food. This kind of plasticity is a valuable characteristic, which has evolved like any other and is under genetic control. The mechanisms of developmental homeostasis control the dynamic interaction between the genotype and the environment so that the phenotype is enabled to survive changes in the environment during its growth.

Homeostasis of the Population

Just as the physiological and behavioral mechanisms that we have discussed help to maintain a steady state in each individual organism during its lifetime, so are there mechanisms that maintain the steady state of the whole population over long periods of time, keeping an equilibrium between it and the environment.

This point was strikingly made by the great paleontologist A. S. Romer who argued convincingly that the evolution of adaptations to land in the form of lungs and legs were really adaptations to remaining in water. If a river or pool dried up, a lungfish could survive the dry spell and an amphibian that could walk downstream to another pool to regain its aqueous habitat would survive, while lungless fish would die. As Romer wrote: "True land life seems to have been only the result of a happy accident." Lungs evolved to survive drought and limbs evolved to reach the water and retain an aqueous lifestyle, not to leave it.

The environment is always oscillating as well as undergoing long-term change. At the same time, natural variation in individuals from generation to generation will tend to cause random changes in the average characteristics of a population. Living populations are homeostatic in the face of both these kinds of variation, insofar as they remain in a relatively steady state in spite of minor environmental and genetic variations. This phenomenon has been termed *genetic homeostasis*. Such a mechanism will maintain an equilibrium between the population and the environment and will thus tend to ensure the survival of the population. In the absence of environmental change, gene frequencies will remain static. The balance between different alleles (the differing states of any given gene) will not alter statistically. This is the *Hardy-Weinberg Law*. Long-term environmental change, however, will result in a homeostatic readjustment on the part of successive generations of the population. This is evolutionary change.

For example, the fittest mean weight and size of a species in a more or less constant environment will be maintained in the face of individual variation. Natural selection will cut down the reproductive rate of the more extreme variants and will thus operate to regulate this characteristic. On the other hand, a gross change in external conditions (such as greater food supply resulting from climatic change) may alter what is the fittest weight, and so a homeostatic adjustment will be made in the population to a new, increased mean weight and size. Such an adjustment constitutes evolution and fulfills the definition of evolution as a *systematic change in the genetic structure of the population*. Such a change is termed an *adaptation* an adaptation to changed environmental conditions. This, then, is evolution by natural selection: shifts in the mean of existing variable characteristics (adaptations), that help to maintain the continuing survival of the population. New mutations are not necessary for such changes, but existing variability is essential for change and so for survival.

As we have seen there are rare circumstances in which the stabilizing pressure of natural selection fails and permits random variations in the genetic structure of the population to become fixed. This *genetic drift* (though it is random and nondirectional) may result in the appearance of neutral or even maladaptive characteristics in the phenotype. Drift may occur when a small population is isolated in a relatively rich environment. Under these conditions novel, but random, characteristics may survive for a time because selection pressures are low. Such a situation is believed to have occurred when a few animals or plants have entered a new environment, perhaps by drifting on a natural raft to an island where they find much nourishment and meet few, if any, predators. Thus there are generally low selection pressures operating. Such a situation may have occurred rarely during the course of primate evolution.

We see, then, that at the level of both the individual phenotype and the population, homeostatic mechanisms are operating to maintain a steady state. In both cases the steady state is the key to survival, and the maintenance by evolution of such effective homeostatic systems is the unique property of living organisms. Even evolutionary change (although it is usually irreversible) can be seen as an attempt to maintain the *status quo* in the face of environmental change. As Romer pointed out, the first adaptations to terrestrial life were no more than an attempt by crossopterygian fish to remain in water, by moving from pool to pool, or from stream to stream. Species are not driven to evolve, but evolution occurs as a product of their survival mechanisms.

We can also see that the individual phenotype is maintained by the mechanism of genetic homeostasis i.e., by the properties of the genes, and of course the function of the phenotype is to ensure survival of the genes. Phenotypes are survival machines for genes. Many different kinds of phenotype have in fact survived, and it will be of interest to consider whether any one form of survival machine is in any way better or more progressive than another.

Behavior

The study of behavior refers to what animals do in their interaction with the environment, and it is concerned to a great extent with the activity of locomotion. Broadly, it is concerned with how animals come into contact with their environment how they breathe, how they touch and move upon the ground, how they raise their young, how they eat the portion of the environment that constitutes food, how they escape from the portion that constitutes predators, and how they communicate with and copulate with the portion that constitutes their own species. Behavioral scientists describe the way that an animal is related to the environment, and much of the interaction between the phenotype and the environment occurs through the medium of behavior. An understanding of the behavior of an animal is therefore necessary for a full understanding of the interaction of a population and its environment; such interaction must be studied in an analysis of the process of evolution. This is a difficult subject, but some broad generalizations may be made as a basis for discussion of the evolution of human behavior. In the first place, behavior, like morphology, arises as a result of the interaction between the genotype and the environment. The potentialities of all behavior patterns are genetically determined, but in the absence of a suitable environment a behavior pattern may not mature, just as an individual may not. Behavior that is derived from information coded in the genotype, with little contribution from the environment, is often called *innate* and is well exemplified among lower animals such as insects. A remarkable example involving a visual "language" is found in the honeybee. By a skillful piece of research conducted over many years, von Frisch and his successors have shown conclusively that a worker bee that has discovered a new source of food will, on returning to the hive, perform a dance on the face of the honeycomb. This dance has been shown to transmit information to other worker bees about the direction, distance, and nature of the food source and is a rare example of a descriptive language among animals. It is significant that it is found among one of the insects with a highly organized society. This descriptive information is in condensed, coded form and accurately enables other bees to find the flowers described. The bees' capacity for this complex behavior pattern is innate, but, since it records variable information, is determined in its details by the conditions of the external environment.

There is a great deal of variation among different species of animals as to the proportion of information input (which determines behavior) that comes from the genotype and the proportion that comes from the environment. There is an almost equally great variation in this factor within each individual species. Although some behaviors are described as innate, none can appear in the total absence of appropriate environmental input. For example, the nipple-searching and sucking reflexes of newborn babies are innate and genetically coded, but only appear as a response to contact with the mother's breast. At the other end of the scale, even behaviors that seem to depend solely on learning are ultimately based on potentials that are coded in the genotype. The ability to play a musical instrument, while learned, is based on a genetic potential not evenly distributed among the population.

Thus, if we consider the genetic or innate component only, we see that it can vary from a detailed, programmed motor response (such as sucking) to a very generalized potential such as musical ability. An instance of an even more generalized potential is car driving. Here the innate component lies in the physical capability of the individual and the possibility of rapid coordination between eye, hand, and foot; as we shall see, all these characteristics are associated with arboreal primates.

The environmental component also varies. It merely consists of the minimal conditions for life and an appropriate stimulus: in our example, the skin of the mother's breast. At the right hand side of the chart, the information input may be not only from the general environment but from other members of the social group, and in this case it may take on the quality of *learning*. However, only a proportion of the input from the environment can be called learning.

The environment influences each behavioral activity anew every time it occurs, obviously a necessity in the case of the bee dance in order that a correct report be given of each fresh discovery. But among many other animals it is clear that the effect of the environment upon each action may be recorded in some way in the central nervous system so as to influence future action. In the latter case some degree of learning has occurred.

In practice, therefore, it is impossible to draw a line between innate and learned behavior. Even the flatworms, with a much simpler internal organization than insects, have been shown to be capable of benefitting from experience; but the more complex the animal is, the greater the part that learning appears to play in determining behavior. Among birds, for example, there is wide variation in the relative contribution of learning and inheritance in the development of the mature song repertory. Some birds (the cuckoo is a classic example) develop their songs quite normally when reared away from their own species, while others develop abnormal songs when reared in this way for they depend on learning, for the proper development of their song. But, most commonly, broad characteristics are inherited and the details are learned.

Such data have been gathered by separating birds from their kind. This is an experiment that cannot be performed easily with mammals because so much behavior is learned that growth is deeply disturbed in the absence of a mother, and it is impossible to isolate the effect of deprivation upon any particular behavior pattern. But, even in the absence of relevant experimental data, it is clear that in nearly all the behavior of the higher mammals there is an element of learning, and it may be very great. Genes supply the capacity for behavior of different kinds, together with some basic components that help form the full behavior pattern; learning makes possible the realization of this full pattern of activity. For example, studies of bird song suggest that in many species genes make possible the capacity to sing and, especially, to sing certain notes and trills; the actual pattern of the song is, however, usually learned.

Clearly, behaviors that are primarily determined by information from the genotype and those primarily determined by information from the environment each have certain advantages. Behavior that is mainly innately generated obviates the need for learning processes that take time and may involve danger, and saves energy on the part of the animal. On the other hand, it is not flexible, and in a fast-changing environment or a long-lived species could be disastrous, since it could be changed only by the slow processes of variation and natural selection. Learned behavior, on the other hand, is expensive in time and danger, but it is much more flexible and allows readaptation to occur throughout a lifetime and in every generation. Just as evolution results from the interaction of factors causing variability and stability, so behavior has this dual character. Heredity can make innate behavior patterns more or less stable (though subject to evolution), while learning introduces a flexibility that may be of great survival value.

There are three different ways of learning: by trial and error, by observation and imitation, and by instruction. The trial-and-error method is found in all animals that can learn and is the sole means of learning in invertebrate animals. Compared with other methods, it is lengthy and dangerous. It is the method usually referred to in discussions of learning and in experiments testing learning ability. For example, rats can learn their way through a maze by trial and error (just as humans can) so long as some reward is offered. The rate at which learning can take place by this method has been used as a rough indication of the degree of complexity evolved in the central nervous system. Learning by trial and error reaches its highest development in mammals. As a method it is highly flexible and adaptable to environmental change, and therefore it has been evolved to a remarkable extent among some groups of mammals, particularly among the primates.

Learning by observation and imitation is, however, a speedier and more sophisticated means of building up behavior patterns. Imitation depends in the first place on the ability of an animal to recognize and copy another member of its species, usually its mother. Imitation is a shortcut to learned behavior, but it introduces a certain inflexibility in the behavior pattern that is not found in trial-and-error learning and is therefore more valuable when the behavior relates to the more constant features of the environment. Rats can learn to pass through mazes more quickly than by trial and error alone if they are allowed into the maze with other rats already trained in the art. Many birds learn their song repertories to a great extent by imitation. Monkeys and apes also learn a great deal that way, and children learn still more and faster. During the early stages of human development, imitation is probably the most important factor in developing behavior patterns.

Finally, instruction is a uniquely human way of creating behavior patterns, for it involves conscious thought and intent. It is the means of learning that is essential to the development of culture, yet is dependent upon culture for its existence.

Another and very different determinant of behavior that has not yet been mentioned may be termed "intelligent thought." It is in all probability a uniquely human characteristic and so is a novelty of great importance in human evolution. Action determined by intelligent thought may draw on learning but is itself not the result of a learned behavior pattern; it is original.

With its three determinants heredity, learning, and intelligence the study of behavior is highly relevant to the study of human evolution, for the evolution of behavior itself has played a vital role in our evolution. In the following sections we shall see more clearly the importance of the less easily detected changes in learning and behavior; we must study them as well as the more obvious changes in anatomy and physiology that characterize our evolution.

Sociobiology and Human Behavior

Sociobiology, the science of the biology of *social behavior*, is an important and growing branch of biology that is of particular interest to us since we are a very social species. The science, with its roots in the study of social insects, is concerned with the behavior of all social animals from insects to humans. Its origin as a new science can be traced to the concept of *inclusive fitness*, first discussed by W. D. Hamilton in 1964.

The concept implies that the "Darwinian fitness" of an individual not only includes its own reproductive success but that of its relatives, which share the individual's genes. It is, as it were, the particular packages of genes that are being selected rather than particular individuals. This became clear from the study of social insects such as bees. In this group of insects, because of their unusual genetics with haploid males (i.e., males carry only one-half the normal diploid gene complement, so that the nucleus of the cells does not undergo meiosis in the production of gametes), the daughters inherit a full set of their father's genes as compared with the usual 50% of their diploid mother's genes. This means that since each worker bee shares (on average) 75% of its genes with its siblings (50% from having the same father and 25% from having the same mother), it pays it to work for the survival of its siblings as much as (in fact more than) for its own putative offspring, were it to have any, who would carry only 50% of its genes.

Similarly, if a bird giving a warning cry can thus protect a number of individuals that share its genes, the gain in survival of its genes may be greater than the loss that would occur if it was killed by a predator. In practice, the selective advantage of an individual assisting a relative in this or any other way will depend on three factors: the risk to its own fitness, the benefit to the relative or relatives, and the degree of relatedness of each relative. We find that on this basis we can account for the selection of what looks like altruistic behavior in close-knit social groups where individuals are closely related.

In reality, however, the behavior is not truly altruistic insofar as an individual is protecting its "own" genes. The phenomenon has been termed *kin selection* or *gene selection*. It is not, however, an alternative to natural selection but an extension of it. The term *gene selection* is appropriate in one sense, but it should be recalled that selection always operates on the phenotype and not on the genotype. The best way to describe this phenomenon, therefore, is to see it as a development of the concept of Darwinian fitness that is now extended to include not only the reproductive success of an individual and its offspring but the reproductive success of all individuals in a social group that carry that particular individual's genes. We shall follow the suggestion of Hamilton and refer to the phenomenon as "inclusive fitness."

The concept is extremely important in understanding the behavior of individuals in the kinds of social groups that are found among the higher primates. We must conclude that natural selection produces individuals that are adapted to maximize the spread of their own genes through their own survival, that of their offspring and their relatives.

This limited kind of *bioaltruism* towards relatives, which is effectively an extension of maternal care, finds a sound genetic basis in the concept of inclusive fitness. The theory can be extended to include a considerable range of behavior, including behavior that has always been considered typically and uniquely human and that would not appear to have a genetic basis. Such behavior would include human altruism, art, love, religion, and so on. E. O. Wilson, who has most strikingly developed the concepts of sociobiology, has attempted to apply them to human evolution and behavior in some detail. The basis for Wilson's view lies in the fact of the ultimate *genetic* control of human populations a control that we cannot escape. He points out that if the brain evolved by natural selection, and we believe that it did, then its capacities to develop and select particular esthetic judgments and religious beliefs must have arisen by the same mechanistic process. They are either direct adaptations to past environments, or at most constructions thrown up secondarily by deeper, less visible activities that were once, long ago, adaptive in a strictly biological sense. "The brain exists in its present form because it promotes the survival and multiplication of the genes that direct its assembly. The human mind is a device for survival and reproduction, and reason is just one of its various techniques". As evolutionary biologists we must accept this general statement, but we are entitled to ask whether it precludes the possibility of a sizable amount of nonadaptive behavior. We need to be clear as to whether all humankind's achievements are part of our biological adaptation, with no other goal, or whether we are now in a position to develop behavior with another (higher?) goal in view a purpose beyond the imperatives created by our genetic history.

Wilson answers this question negatively. He writes: "Genes hold culture on a leash. The leash is very long, but inevitably values will be constrained in accordance with their effects on the human gene pool. Human behavior like the deepest capacities for emotional response which drive and guide it is the circuitous technique by which human genetic material has been and will be kept intact. Morality has no other demonstrable ultimate function".

"The leash is very long." This is surely the critical statement. In fact we now know just how long it is. It is long enough for the human individual to choose to have no children, and for the species to destroy itself entirely, with nuclear weapons. If this were to occur, even in error, and the possibility exists, then we should have escaped the genetic leash completely and our brains would finally have destroyed our genes.

If this is even a possibility, and it is, it demonstrates that human behavior can indeed escape the biological constraints a biologist might predict would limit its wilder excesses. The possibility of self-destruction by a rational species suggests that the genetic leash is so long that humans can and often do act in a nonadaptive manner, and it suggests that they may well be capable of setting goals for themselves that are not adaptive in any biological sense and moving toward them.

Among animals, pleasure is a reward for biologically adaptive behavior (e.g., feeding, sex). Humans have discovered how to generate pleasure without necessarily producing adaptive behavior. Sex with the pill is one example, while listening to Mozart is another. Humans can also set goals for themselves (zero population growth, nirvana, and so on) that seem to fly in the face of biological imperatives. Perhaps they do not, perhaps they themselves are quite complex adaptations, and zero population growth is a possible example of that. However, the conclusion we must face is that we are not driven to look for the adaptive value of every part of human behavior. While we may be satisfied to find it (for example, if homosexuality proved to be a birth control mechanism that arises in conditions of overcrowding), we should not be surprised if we do not. The leash is long enough to permit the most startling freedom for humans, who, if they understand what they are doing, can, with their technology, cut the leash that binds them to their past.

This is possible because the species has evolved the capability of rational thought, which, though it gave great power of adaptation to a technological species, also brought an understanding and made possible a freedom of action that animals do not possess. The brain has given us freedom from the control of the lower brain centers, the limbic system, which makes animals do what they have to do. It has enabled us to escape from the prison of our biology, our past, and given us increasing freedom to explore ourselves and our world. This development, which has done as much as anything else to make humankind the extraordinary species that it is, has been possible because of the success of the human adaptation. Because we are many and widespread, we can afford, as a species, to take risks; we can afford to try out our newfound freedom, and risk our genes in living out our dreams. The ultimate mystery lies, perhaps, in the origin of those dreams: the new goals that give meaning to the lives of so many people.

Evolution and Progress

Evolution, as indicated by the foregoing discussion, is an adjustment in the mean characteristics of a population due to natural selection, which usually results in an increased likelihood of the population's survival in its particular environment. There is no force operating within individuals that causes evolution, nor are systematic changes in the genetic structure of a population necessarily "advances" on the *status quo*; they are changes which maintain that status. Some changes may prove retrospectively to result in more success than others, depending entirely on the definition of success in a particular context.

The idea of success and progress and the idea of evolution are not unrelated. The idea of progress was hardly in existence 300 years ago for, like evolution, it was, above all, a 19th-century concept. Evolution has formed human kind from the primeval ooze, so it is clearly progressive, at least from our viewpoint. But the view *sub specie aeternitatis* is very different. Here is a process, operating in time, which has produced a whole range of living organisms, from viruses to humans. While there has been, on the one hand, a trend toward size and complexity, the smaller and simpler forms of life have at the same time not just survived but evolved in great variety and numbers. The plant kingdom as a whole has achieved a dominant position without the evolution of a nervous system or musculature. Our greatest competitors in the world today are not other mammals but minute and relatively simple viruses, bacteria, and protozoa. As Inge has written:

> *The progressive species have in many cases flourished for a while and then paid the supreme penalty. The living dreadnoughts of the Saurian age (the dinosaurs) have left us their bones, but no progeny. But the microbes, one of which had the honour of killing Alexander the Great at the age of thirtytwo, and so changing the whole course of history, survive and flourish.*

Evolution as a process can hardly be described as progressive in the usual sense of that word; it is characterized by change (not necessarily involving improvement), the sole positive result of which is survival. It is in that case reasonable and perhaps more valuable to examine the process from the human viewpoint in order to discover whether the human species is in any sense progressive compared with the rest of the animal world. In what sense are we more progressive than other animals?

The success of a species could be measured in terms of its numbers, or the gross weight of its biomass (the total mass of all of its individual members); it could be judged in terms of its stability in time, its variety of adaptations, or its geographical range. We humans are in fact actually increasing our population and our total biomass faster than any other organism, yet that can hardly be called a sign of real progress, since we know it to be both undesirable and dangerous. Other species have had population explosions; so even if we excel in such characteristics, they do not explain our special evolutionary position. If we look at *Homo sapiens*, we see, for example, that the species is very complex (especially the nervous system) and self-conscious (which is probably unique in the organic world). It has also intentionally altered its environment. Again, we cannot be sure that it is necessarily progressive to be complex, or to be self-conscious, or to alter the environment. Any of those characteristics could prove disastrous to their possessor, and they may do so to humankind. But their combination does reflect one single evolutionary trend in which humans appear to be progressive: they have an improved homeostatic response to environmental change, or, as Herrick puts it, progressive evolutionary change is "increase in the range and variety of adjustments of the organism to its environment".

Let us examine the mechanism of homeostasis further in order to discover how such a progressive improvement may have come about. A homeostatic system consists basically of three parts: (1) the receptors, which measure the status of critical variables in the internal or external environment; (2) the controller (regulator or mediating mechanism), which acts when the signal from the receptor indicates a need for an adjustment (again, in our example, it is the hypothalamus); and (3) the effector, which the controller uses to bring about correction of a deficiency or excess.

Receptors that sense the status of the internal environment are of course internal, and they are usually found in the brain itself. The survival value of response to changes in the external environment (for example, absence of food) has resulted in the evolution of the sense receptors (which in our example can detect food chemically or visually). The evolution of such receptors makes possible response to the external environment, but by themselves the sense organs may not be sufficiently accurate as generators of input to allow very precise homeostatic adjustment. For example, the very simple eyespot of certain single-celled animals is able to detect the direction of light sources, but it required not just the eye but the vertebrate visual system as a whole to recognize distinct shapes within the overall visual pattern of the environment. Thus, the evolution of the sense organs must be accompanied by the evolution of analytic mechanisms in the brain that can differentiate between different intensities and patterns of stimulus. The evolution of such a refined input mechanism has been accompanied by the evolution of more sophisticated controllers and effectors that generate a wide range of behavioral response.

With the early mammals we find not only the full range of sense organs that we ourselves have, which respond to changes in the external environment, but also the full range of internal homeostatic mechanisms already mentioned and described by Cannon. But during the more recent stages of human evolution, during the primate stage, we find the development of a general-purpose controller the cerebral cortex of the brain and a general-purpose effector the versatile limbs which make possible a wider range of behavior. In this study of human evolution we shall place great emphasis on this new kind of adaptive response, which is less specific and more generalized and versatile. It takes the outward form of learned and intelligent behavior.

Thus, we shall see in primate and human evolution that increased homeostatic response is based on (1) broad sense reception with very detailed analysis of input, (2) an advanced general-purpose computer and controller, the cerebral cortex, and (3) much increased versatility of the effectors, the nerves, muscles, and bones by which the limbs and body move.

It is in this respect that humans can claim to be more progressive than other animals: our sensory inputs supply far more data about the environment than do those of other animals, and our effector organs make possible a greater range of behavior. The evolution of the large brain is the necessary concomitant of this extended interaction with the environment.

How do we reconcile the recognition of progress in human evolution with our earlier definition of evolution as no more than adaptive change? The answer is that evolution involves two kinds of adaptation. The first and simplest is an adjustment by existing homeostatic mechanisms (different mean weight, different salt concentration, different coloration). The second, and the more complex, is an alteration in the functional homeostatic systems themselves. For example, in the evolution of many internal parasites, such as the tapeworm, various functional homeostatic mechanisms have been lost because their external environment (the body of another animal) is more constant than that in which the tapeworm originally evolved as a free-living organism. In this example evolution is not progressive, but the animal has become very highly specialized to exist in a very limited environment.

On the other hand, the evolution of the temperature-regulating mechanism in primitive mammals was something altogether new and certainly progressive, since it gave mammals improved homeostasis and a greater independence from environmental change. Since the appearance of mammals, further improvement in homeostasis has arisen from behavioral rather than physiological mechanisms. Man's body in its structure and function is very similar to that of other mammals; his difference from them lies in his behavior and the mechanisms determining it. Such differences are far-reaching and constitute a progressive change in an evolutionary lineage.

Specialization and Pedomorphosis

The extent to which a population is committed to a particular environment is termed its degree of *specialization*. As we shall see, some primates are more highly adapted to arboreal life, are more specialized in this respect, than others. Thus, their degree of specialization can be seen as their degree of irreversible commitment to their environment: highly specialized species may not survive environmental changes that less specialized species might. Specialization in fact seems to reduce the possibility of further evolution, and, in the long view, highly specialized groups (such as the parasitic tapeworm) may all too often be doomed to ultimate extinction when the host species becomes extinct. From the point of view of survival, therefore, a species must remain fairly *generalized* in its overall character; extreme specialization may ultimately involve extinction and indeed it usually has.

There is, however, one remarkable way out of the narrow corner of overspecialization, and from our position of hindsight we can see that it has occurred on a number of occasions. The process, called *progenesis*, can effect despecialization and operates somewhat as follows: we can suppose, as we have seen, that the adult form of an individual is determined by two sets of genes; structural genes determine the adult characteristics themselves, and regulatory genes determine the rate of growth and time of onset of sexual maturity. Since specializations must be present in the adult stage, they usually appear during growth, and their structure at all stages is determined by the first set of genes. The second set of genes, however, establishes the permanent adult form of the individual by their power to time the onset of sexual maturity. By speeding up the onset of sexual maturity, the adult form may become established and growth may stop before all the specializations can be fully developed. The animal becomes sexually mature when in the young stage; the result is a very remarkable evolutionary change, which is due to a very simple alteration in growth pattern. The phenomenon, discussed by Gould, has been an important factor in evolution. Perhaps the most famous instance is the evolution of the chordates (the ancestral group of the vertebrates) from the echinoderms (the sea urchins and starfish). We have evidence that the primitive chordate evolved from a creature like the larva of a sea urchin that had become sexually mature. There is no doubt that the adult echinoderms are a very unusual and specialized group that could not possibly evolve into chordates; yet by the process of progenesis that specialization is believed to have been cast off, with the result that there appeared what amounted to an altogether different kind of animal of unspecialized form the sexually mature echinoderm larva.

Thus, progenesis gives evolving groups a way out of the dead end of overspecialization, and, with limited mutations affecting mainly growth rate and time of maturity, it can result in a saving generalization of form. A somewhat similar process, called *neoteny*, results in the retardation of somatic development of certain organs or structures. As might be expected, there is evidence that such a development is a frequent evolutionary adaptation and that it has played a part in the evolution of *Homo sapiens*. While the process in no sense accounts for human evolution as a whole, there is no doubt that many characteristics considered typically human are present in the early stages of the development of other primates and that they have been retained in humans by retardation in growth rate of certain parts of the body. For example, the limited body hair of *Homo sapiens* and the fair skin of certain races may be the result of the arrested development of the hair follicles and melanin cells respectively. A list of human neotenous characteristics is given by Montagu, who discusses this interesting phenomenon. The results of both progenesis and neoteny are termed *pedomorphosis* the retention into adult life of juvenile characteristics.

All evolution is due to developmental and genetic changes, and the phenomenon of pedomorphosis is of particular interest because it is a shortcut to drastic change through change in growth rate, without change in structural genes. Thus, it may have come about that although we are in many ways profoundly different from chimpanzees, we nevertheless share with them some 98.4% of our DNA sequences. That 1.6% in which we differ includes, no doubt, not only structural genes but many regulatory genes that control rates of development of the components of the body.

It is also likely that mutations affecting growth rate and the timing of the onset of sexual maturity are less likely to encounter the effect of internal selection than are those affecting structure, because the former threaten less fundamentally the harmony of the whole organism. Though the phenomena of progenesis and neoteny both give evolving species a way out of the trap of overspecialization, they do not require us to change our ideas about the action of natural selection and the way that it brings about evolutionary change.

Organism and Evolution

The organism can be described as a self-regulating, self-reproducing open system that passes vital information (coded in DNA molecules) to copies of itself. It can be considered to be of two parts: the genotype, which consists of the informational nuclear molecules of DNA within each cell of the individual, and the phenotype, which is the whole individual formed by the genotype through its interaction with the environment. It is the genes' survival machine.

The genotype interacts with the environment through a causal chain of which the phenotype and its behavior are the connecting links. This interaction maintains the homeostatic equilibria of the individual and population that are necessary for survival. The individual phenotype maintains the equilibrium of the internal environment for the preservation of the genotype.

Reproduction by the fusion of gametes from different individuals of two sexes creates, over a period of time, a network of sexual relations between individuals that constitutes the interbreeding population. In the formation of the gametes, variation arises in the information store of the DNA, and thus gives variability to the population and makes possible changes in it. Both the extent of variation at a given moment (which is ultimately generated by mutation) and the long-term alterations in the gene pool (which constitute evolution) are controlled by natural selection, which is the process whereby the reproductive capacity of individuals not well adapted to the environment is depleted by environmental interference. The change in the population brought about by changes in the environment (which constitute selection pressure) can be described as adjustments made in the population that serve to maintain a homeostatic equilibrium between it and the environment. Because living processes are so unstable and delicate, such equilibria must be maintained at all times for the survival of life.

Variability between populations in different geographical regions may lead to speciation if gene flow is interrupted by geographical barriers for a certain period of time. Isolation allows for the evolution of morphological and behavioral divergence, which may eventually replace the geographical barriers as an effective isolating mechanism so that genetically discrete species are evolved.

Progress in human evolution is seen as the appearance of new homeostatic mechanisms that make possible survival in the face of ever greater environmental variation. The mechanisms themselves increase in turn the complexity of the phenotype. Progressive evolution in any but an anthropocentric sense cannot be recognized. Evolution is merely a homeostatic adjustment in the mean or average characteristics of a living system in response to environmental change and can be the outcome only of such adjustment and nothing more. By a study of the anatomy, physiology, behavior, and environment of living organisms related to ourselves, it is possible to discuss how human nature was formed and to attempt to understand how each characteristic of our species serves its survival.

3

Origin of the First Organisms

Introduction

The world of living things is made up of a tremendous variety of animals, plants and microbes which are widely different from one another. Nevertheless, among all this diversity, even a person without scientific experience will notice something common to all living things, something which relates them to one another and distinguishes even the very simplest organism from objects belonging to the inorganic world. This direct, and sometimes even unconscious, assumption of the ordinary man concerning the world around him itself contains the most primitive as well as the most general definition of life.

The age-long philosophical quarrels and acrimonious differences of opinion on this subject are fundamentally simply concerned with the question as to what is the essence of this 'something' the essence of life. The idealists see it as something spiritual, the essential nature of which is inaccessible to experimental study, while, according to the materialists, life, like everything else in the world, is material in nature and an explanation of it does not call for the acknowledgement of anything supernatural.

Quite large numbers of scientists now take the view that an understanding of life in general involves no more than a very thorough knowledge of physics and chemistry and a very thorough explanation of all vital phenomena in terms of physical and chemical processes. According to this view there are no specifically biological laws, and the rules which prevail in the inorganic world also govern all the phenomena taking place in living organisms. But this amounts to denying all the essential differences between organisms and the objects of the inorganic world, which is fundamentally unsound. Certainly life is material in nature, but it is not inherent in every sort of material. It is a manifestation of a special form of motion which we only find in organisms and which is absent from the objects of the inorganic world. This form of the motion of matter, in addition to obeying the general physical and chemical laws, also has its own specific laws. If one is to understand life it is therefore important to take into account these qualitative differences from other forms of motion.

Outstanding scientists and thinkers of past centuries and of the present time have formulated numerous definitions of life which, to a greater or lesser extent, indicate what is specific to it.

We cannot discuss all these definitions in any detail and shall here confine ourselves to one which was given by F. Engels as early as the end of the nineteenth century but which still remains extremely pertinent. "Life is the mode of existence of albuminous substances and this mode of existence essentially consists in the constant self-renewal of the chemical constituents of these substances."

Thus Engels characterises albuminous substances as the material bearers of life, and metabolism as their essential function; from this all the rest of the most general attributes of life may be derived. In doing so we must not, as was pointed out in Chapter VI, identify the 'albuminous substances' referred to by Engels with the individual proteins which can now be isolated from living organisms.

Nevertheless, such unjustifiable identification has formed the basis of several attempts in the recent literature to interpret as metabolism the reactions observed by a number of authors, in which amino acids containing isotopically labelled atoms are incorporated into isolated proteins, sometimes involving the substitution of amino acid radicals within the protein molecule. Such an interpretation clearly derives from a confusion between two completely different concepts: (1) biological metabolism in the sense which was described in the previous chapter, i.e. the orderly sequence of processes which seems to be the prerequisite for the existence of any living thing, and (2) 'exchange reactions' or substitution reactions in the purely chemical sense, i.e. phenomena in the course of which two molecules of organic substances or even inorganic salts exchange their atomic groups, e.g.:

$$CH_3COOC_2H_5 + HOC_5H_{11} \rightarrow CH_3COOC_5H_{11} + HOC_2H_5$$

Undoubtedly the molecules of any protein, with their extensive chemical potentialities, can take part in such substitution reactions, but, unlike biological metabolism, these reactions are certainly not absolutely necessary to the existence of the protein molecule. It is well known that individual proteins isolated from living things can be kept under suitable conditions in the native state without any such reactions taking place, while an increase in their ability to enter into such reactions may be an indication of denaturation of the protein.

In contrast to this, metabolism is certainly a necessary condition for the existence of protoplasm or 'albuminous substances' in the sense in which Engels understood the term.

Engels wrote:

> *From the moment when this uninterrupted metamorphosis of its constituents, this constant alternation of nutrition and excretion, no longer takes place in an albuminous body, from that moment the albuminous body itself comes to an end and decomposes, that is, dies.*

It follows from the very theory of open systems' that the continuous renewal of the component parts is a necessary condition for the existence of such systems; that as soon as the flow of water from the tap, or, in the hydrodynamic model, the admission of water, ceases then the system itself immediately ceases to exist as such. Similar considerations apply to any chemical open system.

Principle of Selection

However, living things differ fundamentally from all such open systems in the orderly regulation of their metabolism and the 'purposefulness' of their internal structure. Not only are the many tens and hundreds of thousands of chemical reactions which occur in protoplasm, and which together constitute its metabolism, strictly co-ordinated with one another in time, harmoniously composed into a single series of processes which constantly repeat themselves, but the whole series is directed towards a single goal, towards the uninterrupted self-preservation and self-reproduction of the living system as a whole in accordance with the conditions of the surrounding medium.

Here the term 'purposefulness' should, of course, not be taken in an idealist sense as the 'fulfilment of some higher design'. The word is used to denote the appropriateness of the organisation of the whole system to its self-preservation and self-reproduction, and also to denote adaptation of the structure of its separate parts to the most efficient and co-ordinated fulfilment of those functions necessary to life which these parts subserve in the system as a whole.

The high degree of adaptation of the separate organs to the carrying out of their functions and the general 'purposefulness' of the whole organisation are extremely evident even from a superficial knowledge of higher living things. They were noticed by mankind a very long time ago and were expressed by the 'entelechy' of Aristotle. The essential nature of this purposefulness appeared to be mystical and supernatural until Darwin gave a rational and materialistic explanation of the way in which this 'purposefulness' could arise as a result of natural selection.

However, 'purposefulness' of structure is not confined to higher beings, it extends downwards through the whole world of living things, right to the bottom, to the most elementary forms of life. It is essential for any living body but is absent from the objects of the inorganic world. The only exceptions are machines, but the purposefulness of their structure, their adaptation to the performance of particular tasks, is determined by the creative intention of those who build them. Machines cannot arise of their own accord by purely physical and chemical means. It is therefore pointless to seek an explanation of them in purely physical and chemical terms. The origin of the organisation of protoplasm which characterises living organisms, biological metabolism, is understandable only on the basis of the same principles which govern the origin of the 'purposefulness' of the structure of higher organisms, that is to say, on the basis of the interaction between the organism and the environment and on the basis of the Darwinian principle of natural selection. This new biological law arose during the actual process of the establishment of life and later took a leading part in the development of all living matter.

But can this law be applied to any system other than the living organism? As we have already seen the attempt to apply the principle of natural selection to the evolution of separate molecules cannot be held to be satisfactory. However, we shall adopt a different approach if we try to imagine the possibility of the evolution of those systems which we postulated in the previous chapter as being the starting point on the road to the development of living systems, that is, to the evolution of the drops of complex coacervates which have the properties of open systems and the network of interdependent reactions characteristic of such systems.

In the very origin of such individual multimolecular formations there was already inherent the necessity for their further progressive development. During the time when organic material was completely merged with its environment, while it was dissolved in the waters of the primaeval seas and oceans, its evolution could be considered as a whole. However, as soon as it became concentrated at definite points, in colloidal multimolecular systems, as soon as these formations became separated from the surrounding medium by a more or less clearly defined boundary and attained a certain individuality, new and more complicated conditions were at once created. The later history of any individual coacervate drop might differ substantially from that of another coexistent system. The fate of such a drop depended not only on the general conditions of the external medium, but also on the specific internal organisation in space and time of the system in question. The details of this organisation were peculiar to the particular drop and may have been somewhat different in other drops, each system having its own characteristic peculiarities.

What were the conditions which determined the existence of any coacervate drop in the waters of the primaeval hydrosphere? Complex coacervates obtained artificially, by simply mixing solutions of two differently charged colloids, are, as we have seen, formations with a static stability. The greater or less duration of their existence is determined by the conditions of solubility or the presence of surface membranes and is associated with the maintenance of the constancy of the properties of the system in time.

Thus, in such a coacervate drop, the slower any particular change takes place and the more constant the surrounding medium remains, the greater will be the stability of the system and the less its chance of disappearing as an individual formation during the time it is under observation under the conditions of a laboratory experiment.

This, however, was not the sort of stability manifested by the systems which played the decisive part in the evolution of matter on the way to the origin of life. This evolution could only proceed on the basis of interaction between the systems and the external medium in contact with them, i.e. on the basis of the formation of open systems. We must remember that the coacervate drops, which arose somehow in the primaeval hydrosphere, were immersed, not simply in water, but in a solution of various organic compounds and inorganic salts which were certainly capable of entering into the coacervate drop and interacting chemically with the substances of which it was composed. If we do so it will be clear to us that under these conditions the stability of the drop could not retain its static nature. The drop would, to some extent, assume the character of an open system.

This would occur specially readily when the actual formation of the drop was based on a previous chemical organisation in time like that postulated by M. Sugita. However, let us suppose that the drop arose under purely colloidal conditions, that the whole process of its formation resulted simply from the concentration of protein-like substances and others of high molecular weight at a definite place, and from the formation of a surface membrane separating the collection of these substances from the external medium. Even so the molecules of the external medium must have passed selectively through the surface membrane of the drop or been adsorbed selectively by the compounds contained in it and reacted with them in one way or another, the products of the reaction either being retained within the drop or passing out of it back into the external medium. Although these reactions took place very slowly and did not form an interacting network of processes, and although the conditions necessary for the prolonged existence and stability of the coacervate drop were still not present, nevertheless, even at this primitive stage of evolution of our original systems two circumstances were manifest which were of great importance for the further development of matter.

On the one hand the individual peculiarities of the physico-chemical organisation of each separate coacervate drop imposed a definite pattern on the chemical reactions which took place within that drop. The presence in a given drop of this or that compound or radical, the presence or absence of simple inorganic catalysts such as salts of iron, copper, calcium, etc., the degree of concentration of protein-like substances and other substances of high molecular weight forming the coacervate, its particular structure, all these affected the rate and direction of the various chemical reactions which occurred within the given drop, all these imparted a specific character to the chemical processes which took place within it. Thus there appeared a certain connection between the individual structure and organisation of a given drop and the character of the chemical transformation carried out within it. In other drops these transformations occurred and were co-ordinated in different ways, depending on the peculiarities of each particular drop.

On the other hand, any chemical processes, even uncoordinated ones, occurring within a drop, and, even more, any connected group of processes, could not be without effect on its future. Some of them led to greater stability, to a more prolonged existence of the coacervate system under the conditions prevailing in the external medium.

From this point of view they were advantageous, they were of positive significance. On the other hand, other processes and groups of processes were of a negative character, they were inimical to the particular individual formation, leading to its dissolution, to the disappearance of the drop in which they arose.

However, such coacervate systems cannot have played any essential part in the further evolution of organic formations as their individual history was short and quickly brought to a close. The only systems which maintained themselves in existence for a more or less prolonged period under the conditions prevailing in the external medium were those which had an individual organisation based on chemical reactions which were favourable for their existence.

Thus, even at this stage of the evolution of matter there appeared a certain 'selection' of organised colloidal systems on the basis of the suitability of their organisation to the function of preserving the uninterrupted interaction of the system and the surrounding medium under given circumstances. This 'selection' was, of course, of a very primitive kind and not directly to be compared with fully developed 'natural selection' in the strictly biological sense of the term.

Nevertheless the further evolution of organic systems was controlled by 'selection' of this sort and thus acquired a definite direction.

Processes of Self-renewal of the Systems

In the first place this directed evolution led to an essential alteration in the character of the stability of the original colloidal systems. The stability of the coacervate drops which first arose in the waters of the hydrosphere may originally have been governed by the same static principles which govern the stability of coacervates of gelatin and gum arabic produced artificially in the laboratory.

The coacervate state and the organisation of the processes taking place within the drop may, to some extent, exist independently of one another. However, for reasons which have already been indicated, during the course of directed evolution these two aspects of the organisation must afterwards have become more and more unified within the single system, because the existence of the system depended on a network of reactions carried out within it while, conversely, the network was determined by the organisation of the system as a whole. If the system was not co-ordinated but, nevertheless, interacted with the external medium, it would very quickly disintegrate and disappear as an individual formation. If the interaction between the system and the medium stopped for any reason, then the system would become static and, as such, cease to take part in the general process of evolution.

For example, if the stability of the drop depended on the formation of strong surface layers and if these disintegrated spontaneously at a definite rate but could be built up again in the course of chemical reactions within the drop, then the stability of the drop would depend on the relative rates of disintegration and reconstruction of the surface layers. If the chemical reactions took place fast enough in the drop, with a corresponding fast rate of formation of the firm surface layers, then the dynamic stability of the drop might also be very great. In this case an increase in the rate of the chemical reactions within the drop would have favoured its stability. The increased rate of reaction within a drop would increase its stability, and prolong its survival under such conditions. If the rate of formation of the surface layers became less than the rate of their destruction such a drop would soon disintegrate. Finally, if the surface layers themselves were very strong and stable but not associated with any chemical reactions within the drop, then such static colloidal systems would be excluded from the course of the evolutionary process.

Accordingly, as a result of the directed evolution of the original systems, their stability took on a more and more dynamic character. The coacervate drops were gradually transformed into open systems the very existence of which, under the given conditions of the external medium, depended on the organisation of the processes taking place within them. In other words, there arose systems in which there was a background of continuous processes of self-renewal and which could preserve themselves and exist for a long period on the basis of constant interaction with the external medium. The origin of this capacity for self-preservation may be regarded as the first result of the directed evolution of our original systems.

Origin of the Capacity of the Systems for Self-preservation and Growth

The second step forward in the same direction was the emergence of systems which could not merely preserve themselves, but could also grow, increasing their mass by drawing substances from the external medium. As was pointed out in the previous chapter, the stationary state of open systems is maintained constant, not because the free energy of the system is at a minimum as in thermodynamic equilibrium, but because the system is continually receiving free energy from the surrounding medium in an amount which compensates for the expenditure of free energy within the system. In such chemical open systems as the coacervate drops of the primaeval ocean would seem to have been, the intake of free energy was mainly due to the entry into the drop of organic compounds which were relatively rich in energy and which underwent some sort of chemical reaction within the drop. When chemical reactions are taking place, however, closed and open systems differ from one another in that in the former equilibrium is characterised by the reaction occurring at the same rate in both directions so that, in the nature of things, there can be no increase in mass. In open systems in the stationary state, on the other hand, the rate of the reaction is considerably greater in one direction than in the other, and it follows that there may exist in them a coordination of processes leading to an increase in the mass of the system. Such systems enjoyed an undoubted advantage in the process of directed evolution and therefore, owing to the action of 'selection', they came to occupy a predominant position in the general extension of organised formations.

In the absence of any appropriate experiments, even with models, one cannot say anything definite about the nature of such growth in our original systems. They might have become larger in the form of uniform layers of coacervate, but they might also have become divided into separate drops.

Drops of liquid having a limited mutual solubility or artificial static coacervates cannot divide themselves spontaneously. The forces of surface tension are always tending to make them coalesce and it is only the presence of surface membranes which, to some extent, prevents this from happening. However, as we know, dispersion of this sort may be achieved, even in such static systems, by means of external influences such as simple shaking, which may lead to emulsification.

The dispersion of the primaeval growing coacervate may also have occurred in this way. However, as these were of the nature of dynamic stationary systems the existence of which was bound up with the occurrence of processes within them, their dispersion may have been evoked by internal factors. It may, for instance, have occurred when the osmotic pressure, which was increasing rapidly owing to the hydrolysis of compounds of high molecular weight, became too great for the strength of the surface layer of the drop.

Thus, owing to the constant interaction of our original systems with their environment, there must have occurred a gradual increase in the amount of material organised in the systems. But as this increase always occurred under the influence of 'selection' the only systems which were preserved for further evolution were those which were most highly developed, so that the quality of this organisation was always changing in a particular direction. The systems did not merely become more dynamically stable, they also became more dynamic. We may regard this phenomenon as the third important step in the directed evolution of our original systems on the way to the development of life.

In the first stages of the evolution under consideration, when one could study the fate of isolated coacervate drops without taking into account their relation to other such drops, the factors which were of paramount importance for the prolonged existence of the drop in question as an open system, for its self-preservation under conditions of constant interchange with the surrounding medium, were the relative rates of the processes taking place within it and not the absolute values of these rates.

Origin of the Highly Dynamic State of the Systems

The position is radically altered if we include in the field of our investigations not merely one, but several open systems, existing simultaneously within a particular medium. This may be shown even by working with relatively simple models. For example, when there are several hydrodynamic stationary systems with common initial and final reservoirs, the greater part of the water will proceed through the system which enables it to pass through most quickly. In the case of several parallel, chemical open systems with a common external medium it is obvious that the main flow of substances will pass through the system which, by virtue of its internal organisation (e.g. the presence of more efficient catalysts, etc.), provides the greatest over-all rate of chemical transformation. In this sense the chemical stationary system in which chemical processes occur fastest will have an advantage over other parallel chemical stationary systems so long as the increased rate of occurrence of the processes does not disturb the relationship of rates necessary for the selfpreservation of the system; that is to say, if it is compatible with the prolonged existence of the particular open system. The attainment of the maximal rate by a process involves not merely the acceleration of one of the stages of the transformation but the establishment of a more effective relationship between all the parameters of the process.

From what has been said it is clear that a dynamically stable coacervate drop capable of self-preservation and growth, which had acquired the ability to transform substances more quickly during its interaction with the external medium, would have a significant advantage over other drops which were immersed in the same solution of inorganic and organic compounds but in which the characteristic chemical processes proceeded considerably more slowly. In the general mass of coacervates the relative proportion of such more dynamic drops would become greater and greater. There arose a special kind of competition among the drops, based on the speed with which reactions were accomplished within them and the rate of their growth. For this reason the principle of the greatest speed (which must, nevertheless, be compatible with the existence of the stationary system as such) was a very important factor in the directed evolution of organised formations.

Origin of Systems Capable of Reproducing Themselves

It must, however, be pointed out that the capacity for self-preservation, and even for rapid growth, of the whole dynamic system did not imply the complete immutability of the system. On the contrary, the stationary drop of a coacervate, or any other open system, may be preserved as a whole for a certain time while changing continually in regard to both its composition and the network of processes taking place within it, always assuming that these changes do not disturb its dynamic stability.

Changes of this sort were, in fact, a necessary part of the process of the emergence of life for they guaranteed the evolution of the initial systems. Without these changes no new material would have been provided for selection and the further development of the systems would have been frozen and brought to a standstill at some point.

Naturally, it was of the utmost importance that these changes should not overstep the bounds of the dynamic stability of the systems. Otherwise any markedly unstable compounds which arose would be in constant danger of passing out of equilibrium and disappearing. Therefore, when there was rapid and massive growth of the original systems, selection took place, the only ones which were preserved for further evolution being those in which the network of reactions was so co-ordinated that there arose stationary chains of reactions which were constantly repeated or, even better, closed cycles of reactions in which the reactions always followed the same circle and branching only occurred at definite points on the circle leading to the constantly repeated formation of this or that metabolic product. This constant repetition of connected reactions, co-ordinated in a single network, also led to the emergence of a property characteristic of living things, that of self-reproduction. This may be taken as the origin of life. At this stage in the evolution of matter natural selection assumed its full biological meaning and formed the basis for the faster elaboration of higher and higher degrees of adaptation of living organisms to the conditions under which they existed, of the exact correspondence of all the details of their internal structure to their vital functions. In other words, there appeared that striking 'purposefulness' of the structure of living bodies upon which we have already remarked.

The opinion is fairly widely held in contemporary scientific literature that the capacity for self-reproduction is to be found even in the chemical form of the motion of matter, that it can be a property of isolated molecules. Until comparatively recently many biologists regarded the constant formation of particular substances within the organism as being the result of the presence in the organism of readymade moulds for those particular substances. These moulds were supposed to 'multiply' in some way and thus be responsible for the constancy of the composition and structure of the organism and for its reproduction.

In particular this opinion was once very clearly expounded by N. Kol'tsov. He believed that the formation, not only of proteins, but also of other components of the living cell, such as chlorophyll and the anthocyanins, occurred because the protoplasm already contained corresponding molecules which acted as templates for their formation. These molecules 'multiplied' and thus exactly reproduced their own structures.

The factual evidence of contemporary biochemistry was, however, radically opposed to this opinion and revealed a completely different mechanism of biosynthesis based on the constancy of certain sequences of biochemical reactions. For example, butyric acid is formed by some species of bacteria, not because it was present in them beforehand, but because the sugar which is taken up from the surrounding medium by the bacterial cells is broken down to acetic acid by means of a series of strictly co-ordinated reactions. The acetic acid then combines with coenzyme A and is thus enabled to condense to form aceto-acetic acid which is then reduced to butyric acid.

If the sequence of these reactions were somewhat different, the end products might be alcohol and carbon dioxide, as in yeast, or lactic acid, as in some bacteria. The same principle of the constancy of a definite sequence of reactions is also responsible for the synthesis of the higher fatty acids, amino acids and a whole series of other compounds.

The chemical studies carried out 20 years ago by R. Robinson and the biological work of R. Scott-Moncrieff showed that the anthocyanins mentioned above arise in the same way in plants as a result of the occurrence in them of a definite sequence of reactions of condensation, oxidoreduction, methylation, acetylation, etc. According to the order in which these reactions occur in the petals of different flowers, various derivatives of flavones and anthocyanins are formed, and the particular combinations in which these substances are present give the petals their characteristic colours.

A similar mechanism has also been discovered for the formation of various terpenes in plants from which essential oils are obtained. The terpenes appear in them as a result of sequences of reactions which are specifically determinate for each plant and include polymerisation, hydration, oxidation and ring formation. The same is true for the synthesis of tannins, alkaloids, vitamins and various porphyrin derivatives, chlorophyll in particular. Very detailed evidence has also been obtained recently concerning the biosynthesis of such extremely complicated and specific substances as antibiotics. Here also a definite sequence of chemical transformations is involved.

Thus lactic or butyric acids are formed in particular species of bacteria, nicotine in tobacco plants, tannin in tea leaves, vitamins in yeast cells and streptomycin in actinomycetes, not because pre-formed molecules of these substances were already present in the objects in question, but because, at a given stage in their life cycles, chemical transformations are carried out within them in a definite, co-ordinated sequence. The constancy of the formation of the substances is simply a manifestation of the constancy of the sequences of the reactions. Here there is no 'self-reproduction' of molecules in the literal sense of the term, no multiplication of them; here new molecules of exactly the same kind are repeatedly produced. The sequence of reactions on which this phenomenon is based does not depend on any single individual factor but is a manifestation of the whole organisation of the protoplasm in its relationship to its environment.

The biosynthesis of proteins constitutes no exception in this respect. Attempts to treat it as an autocatalytic process, in which one molecule of a given substance arises as a result of the catalytic activity of another of exactly the same sort which was already present, have recently proved a complete fiasco. The experiments of A. Gierer and

G. Schramm are particularly convincing in this connection. They showed that a single nucleic acid of tobacco mosaic virus completely freed from protein, when introduced into the plant, will evoke the formation in it of a specific protein which was not previously present in the plant. In this case there could be no question of any autocatalysis in the strictly chemical sense of the term. There was only definite co-ordinated interaction of all the processes of the cells of the tobacco leaf, which were somewhat altered in character by the introduction of a new factor, the viral nucleic acid. The nucleic acid as an individual substance, a compound considered in isolation, could certainly not synthesise a protein by itself. It is only effective against the general background of the whole metabolism of the tobacco plant, as is confirmed by all the evidence at present available. The harmonious participation of a long series of catalytic systems is required for the biosynthesis of proteins, some providing the energy needed for the synthesis, some determining the strictly regular and constant relationship between the rates of the different reactions and, finally, some systems which control the spatial organisation of the protein molecule in the process of its synthesis. Among these systems which determine the specific structure of the protein, nucleic acid plays a very important part, but it does not seem to be the sole determinant, it simply constitutes a part of the general organisation of the living system.

As has been pointed out above, nucleic acid itself also arises in the living organism in accordance with the same rules as the other components of the protoplasm, that is to say, on the basis of strictly co-ordinated, constantly repeated, catalytically induced exchange reactions.

It is clear that no substance which forms a major component of protoplasm can be reproduced by a chance or easily attained relationship between the rates of reactions. It requires the absolutely constant, continually repeated chains and cycles of reactions which together comprise the network of the self-reproducing, living, open system. As we have seen above, the origin of such a system may be regarded, theoretically, as a result of the directed evolution of our original, dynamically stable, colloidal formations.

The living systems which were first formed already had all the features needed for their selection to be of the nature of purposeful 'natural selection' in the biological sense of the expression. Further improvements in their internal organisation, rationalisation of their metabolism, therefore, went forward at a faster pace. As a direct result of this, all intermediate forms of organisation were destroyed, swept from the face of the Earth by natural selection. This is why we have now no possibility of studying these forms directly and filling in, with factual material, the abyss which exists between the organisation of the original systems and the organisation of even the simplest of present-day organisms.

The Evolution of Metabolism: The Origin of Enzymes

Experiments with models which reproduce the phenomena in dynamically stable colloidal formations may, perhaps, play an important part in this connection. Studies of this sort are, however, still only beginning to be made and the results obtained from them are still very modest. Therefore, if we wish to formulate any sort of idea concerning the actual forms which developed during the course of evolution from the original systems to the first organisms, we must make as much use as possible of the data of comparative biochemistry (this is done more fully in the next chapter) and the results obtained from a study of the metabolism, or separate aspects of the metabolism, of isolated protoplasmic structures and collections of enzymic systems. In this way we may be able to reveal various features common to all living organisms and may try to form a mental picture of how these features could have arisen during the process of directed evolution of our original systems or in the earliest stages of the development of life.

As we have remarked again and again, the fundamental organisation of living matter is its organisation in time. The phenomena which take place in it in a definite, regular order together constitute its metabolism.

The individual reactions which occur in protoplasm are rather simple and uniform. They are the reactions, familiar to chemists, of oxidation, reduction, hydrolysis, phosphorolysis, aldol condensation, the breaking of carbon-carbon bonds, etc. Any of these may be brought about outside the organism and there is nothing specifically vital about them. What would seem to be specific to living bodies is the definite organisation in time of these reactions in them, to form a single complete system, an abundantly branching network of reactions. In living bodies these reactions do not take place chaotically but bear a strictly determined relationship to one another. The colossal diversity of organic compounds which is to be found in the world of living things does not depend on diversity and complication of the separate individual reactions but on the diversity of their combinations, the variations in the order in which they occur in the different cells of the organism at particular stages of development. This sequence of chemical reactions forms the basis of both the synthesis and the breakdown of the substances of protoplasm. It forms the basis of such vital phenomena as the synthesis of proteins, fermentation, respiration, photosynthesis, etc. In the respiratory and photosynthetic processes sugar and oxygen, carbon dioxide and water, are only the first and last links; between them there are long chains of chemical transformations. In these chains the intermediate product which is produced by one reaction immediately enters into the next reaction, which is strictly determinate for the vital process in question. If these sequences are changed, if any single link in the chain of transformations is removed or altered, then the whole process will become quite different or even be thrown right out of action. As we have seen, these organisational features, which are characteristic of everything living, are exactly analogous, in principle, to the network of chemical transformations which forms the basis of any more or less complicated chemical open system. As in these systems, so in living things, the characteristic order of phenomena which has been described is based on a close co-ordination of the rates of the chemical reactions which form the individual links of the long and labyrinthine chain of metabolism.

Organic substances, which are the essential components of living systems, seem to be the only material which can form the basis of such chains of reactions. It is characteristic of these substances that they can react in the most diverse ways. Although they have tremendous chemical potentialities, these are only realised extremely slowly under ordinary conditions and in isolation. This very slow rate of reaction depends essentially on the great amount of energy of activation, i.e. the high energy-barrier which molecules of organic substances must surmount before they can participate in any chemical reaction. However, depending on all the combinations of circumstances under which any given reaction takes place, its velocity may vary within very wide limits.

If the conditions are such that only one of the reactions possible for any particular organic substance occurs very fast while all the rest of the possible reactions proceed comparatively slowly then, naturally, the practical significance of the latter will be quite negligible in the over-all result. In other words, there lie before each organic substance in protoplasm many routes of chemical transformation which are thermodynamically open to it. In fact, however, each compound which enters the protoplasm from the environment, and any intermediate product which may be formed within the protoplasm, will be changed during metabolism only in the direction in which it can react most quickly. All the rest of the reactions, which take place more slowly, will simply not have time to occur to any significant extent. It is in this way that there are formed those strictly determined chains and cycles of successive quick reactions which together constitute the more or less ramifying network of metabolism.

A simple homogeneous mixture of organic substances, or even a newly formed coacervate drop which has not yet been transformed into a well-organised open system, presents, from this point of view, a very wide but completely untrammelled field of chemical possibilities. The same great difficulties and obstacles hinder movement in any direction in this field. In contrast to this, selection has led to the presence in protoplasm of definite paths of biochemical processes, a whole network of 'rationally built roads' along which there proceeds at a great rate and in 'orderly columns' the chemical transformation of substances and the associated conversion of energy.

This highly-developed order, which depends on the definite relationships between the velocities of the reactions, is regulated in the living body by many factors. The most important of these is the catalytic activity of the enzymes.

Nowadays the study of enzymes, enzymology, has grown into an extensive and independent field of knowledge in which an immeasurable amount of work is being done. Many enzymes have now been isolated from living organisms in the form of highly purified crystalline preparations which have been studied in detail as regards both their chemical nature and the mechanism of their catalytic activity.

These enzymes have been found, without exception, to be simple or conjugated proteins. The prosthetic groups of the latter consist, in most cases, of organometallic compounds or various vitamins. There can now be no doubt that each cell contains a whole collection of diverse enzymes and that the majority of the proteins of the living body have enzymic activity. Thus enzymes would seem to constitute the bulk of the proteins of protoplasm.

The fact that enzymes seem to be chemically proteins, having a definite sequence of amino acid residues in their polypeptide chains and a definite internal structure of their molecules, determines a number of the peculiarities which distinguish enzymes from all other catalysts known to us. The most important of these is their intense catalytic activity.

There are known to be a large number of inorganic and organic substances which can hasten the same reactions as those affected by enzymes, but there is no comparison between the strengths of their catalytic activities. For example, hydrogen ions can catalyse the hydrolytic reaction whereby sucrose is hydrolysed to glucose and fructose, a reaction which is also catalysed by the invertase of yeast, but the enzyme is at least ten million times as effective. The very simple nitrogencontaining organic compound, methylamine, increases the rate of breakdown of pyruvic acid. So does the enzyme carboxylase, but the catalytic activity of the enzyme is about thirty million times as great as that of methylamine. The ferric ion appreciably facilitates the breakdown of hydrogen peroxide into water and oxygen. The enzyme catalase, which is a combination of an iron-porphyrin complex with a specific protein has the same effect but brings about the reaction about 10 times as fast as inorganic iron.

The complicated structure of the protein molecule is also responsible for the second important peculiarity of enzymes, the high specificity of their action. Inorganic catalysts are rather indiscriminate in their action. For example sucrose, maltose, starch, proteins and many other substances may all be hydrolysed equally well by hydrogen ions. But enzymes act in a highly specific way, only catalysing particular reactions. They only break the bonds between certain definite groups of atoms and leave others quite intact, although these may be very similar to those of their substrates. If, therefore, we have any organic substance which is capable of a number of chemical changes, then, in the presence of any one enzyme it will react with remarkable speed, but only in one particular direction. For example, pyruvic acid in the yeast cell, where the enzyme carboxylase is present, is almost entirely broken down to carbon dioxide and acetaldehyde and it is only the acetaldehyde which is reduced to alcohol by the action of a specific dehydrogenase. In the lactic acid bacillus, on the other hand, where there is no carboxylase, pyruvic acid is reduced directly to lactic acid and is not decarboxylated to any considerable extent. Thus the highly specific action of enzymes is a very important factor in the organisation of protoplasm. Less specific catalysts would not have this capacity to determine the direction in which any particular organic substance in the protoplasm would undergo chemical change.

The mechanism of enzymic reactions has now been studied from various points of view by many authors but, so far, the problem cannot be considered to have been solved. In its most general form, the participation of enzymes in metabolism may be presented as follows: The substance which is undergoing the reaction in question (the substrate) first forms a very short-lived intermediate compound with the enzymic protein. This requires a certain correspondence of structure between the enzyme and the substrate. If this is absent no catalysis whatsoever can take place. When this correspondence exists the reaction between the enzyme and the substrate requires considerably less energy of activation and therefore takes place very fast at ordinary temperatures. However, owing to the specific properties of the enzyme molecule, the intermediate, enzyme-substrate, compound is very unstable. It very soon undergoes a further alteration, in the course of which the substrate is changed in the appropriate way and the enzyme is regenerated and can once more form an intermediate compound with a fresh portion of the substrate.

Reactions whereby the substrate is transformed without the help of an enzyme usually require a high energy of activation and therefore take place so slowly that they cannot play a decisive part in metabolism, which is rapid. When the enzyme is present, the high energy barrier seems to be broken down and the route via the intermediate compound seems to be considerably easier and faster.

Thus, in order that any chemical ingredient may actually take part in metabolism, it must first interact with a protein to form a definite intermediate compound. If not, its chemical potentialities will be realised so slowly as to be of no significance in the rapidly flowing process of life. Thus the direction in which any compound is altered in the course of metabolism depends not only on the molecular structure of the compound, but also on the enzymic activity of the protoplasmic proteins with which it becomes involved in the course of metabolism.

Thus, in enzymes, living bodies not only have powerful accelerators of chemical processes, but also an extremely efficient chemical apparatus which can direct these processes along strictly determined channels. This is, in fact, the essential function of enzymes in living bodies, and it must be said that enzymes are extremely efficient 'instruments' for the performance of this function. Their structure is amazingly precisely adapted to the carrying out of this function in the organisms. One has but to make a slight change in the structure of the enzyme complex, to rearrange or block one or other of the chemical groups of its prosthetic part or to disturb the structure of its protein component, and the catalytic activity and specificity of the enzyme are markedly diminished. Thus, even in enzymes, we can already see the suitability of structure to function, the internal 'purposefulness' which is so characteristic of living matter in general.

The study of the formation of enzymes in living bodies, their biosynthesis, is, as yet, really only just beginning; most attention has so far been paid to the question of the 'adaptive' origin of enzymes. We know very little about this matter; it is only clear that the biosynthesis of enzymes, like that of proteins and the other components of protoplasm, must occur by many stages. It is quite unnecessary that, in the course of this biosynthesis, all the elements out of which the enzyme complex is 'assembled' should have been synthesised by one and the same organism. They are very often taken in ready-made from the environment in the form of vitamins or parts of vitamins, essential amino acids, etc.

How could such a highly developed catalytic apparatus have arisen in the first place in the process of the directed evolution of our original systems?

Their great catalytic activity and specificity, was primarily based on the strictly ordered arrangement of atomic groups in their complicatedly constructed molecules. As a result of this, the catalytic activity of each of the groups and radicals is extremely 'advantageously' combined with activating groups which considerably augment their catalytic effects or facilitate the combination of the enzyme with the substrate. This takes place in enzyme proteins, in which such a structure is associated with a definite arrangement of amino acid residues in the polypeptide chain and a definite internal structure of the protein particle as a whole. This is just what may be seen, for example, in the structure of the prosthetic groups of the conjugated-protein enzymes.

We have already mentioned the work of W. Langenbeck on the construction of artificial models of the enzyme carboxylase. In this work the author started from the observation that such a simple compound as methylamine can catalyse the reaction of decarboxylation of pyruvic acid, this catalytic activity being a property of the amino group. The inclusion of a carboxyl group in the methylamine molecule increases its catalytic activity 19-fold, although the carboxyl group itself has no catalytic activity. The catalytic activity of methylamine derivatives may again be increased by the further addition of aromatic and heterocyclic rings. Following this up, Langenbeck finally got a compound (hydroxyaminonaphthoxindole) which had a carboxylase activity 4,000 times as great as that of the original methylamine.

Vitamin B1 forms the prosthetic group of natural carboxylase from yeast. Its molecule, like Langenbeck's models, contains a catalytically active amino group which is combined with two complicated heterocyclic rings. The combination seems to be more effective here than in the artificial model. But it is only when vitamin B1 is combined with a specific protein through a phosphoric group that it acquires the extremely powerful catalytic activity characteristic of the enzyme. Neither the vitamin itself, nor the carboxylase protein, taken alone, have this power and it is only their combination in a special way which gives the enzyme its great activity and specificity.

This sort of structure of carboxylase is a demonstrable instance of one case of the internal organisation of protoplasm. So long as only the separate parts of the enzyme are present or these parts are not combined with one another in a special way, their catalytic activity is small and they carry out their function in the living body badly. If the enzyme is to have its characteristic efficiency in this respect its separate components must be combined together in a special way, but this cannot occur by chance.

Catalase may serve as another analogous example. As we have already pointed out, even ferric ions can catalyse the breakdown of hydrogen peroxide to water and oxygen, but this is only a weak effect. If the iron is combined with a porphyrin nucleus to form haemin, the catalytic activity is increased about a thousand fold. In the natural enzyme, catalase, the haemin is combined with a specific protein and this further increases its catalytic activity many million fold.

In the systems which we postulated as being the starting point for the process of evolution on the way to the origin of life, in coacervate drops having the properties of open systems, the chemical reactions which formed the network of the system must, at first, have occurred very slowly. A certain speeding up of isolated reactions may have been achieved, mainly by means of the catalytic effect of such inorganic salts (e.g. those of calcium, iron, copper and vanadium) as may have been present in large enough quantities in the waters of the primaeval ocean.

Certainly even such a very slight increase in rate must have played a decisive part in the establishment of a definite sequence of reactions, in the organisation of the network of chemical processes in our open systems. In particular, owing to the catalytic activity of inorganic iron, the breakdown of hydrogen peroxide into water and oxygen might thus have occupied a place in the network if it somehow favoured the dynamic stability of the drop, its preservation for a long time or even its growth under the conditions of its interaction with the external medium.

Now, let us suppose that some of these coacervate drops, owing to their adsorptive powers or for other reasons, could take in from the surrounding medium the porphyrins which were formed there by purely abiogenic means, just as some contemporary organisms extract from their environment vitamins which they need for the synthesis of enzymes. On combining with the iron, the porphyrin would markedly increase its catalytic activity, and if such a speeding up of the reaction under discussion was favourable to the dynamic stability of the drops in which it occurred, then these drops would enjoy a considerable advantage in the process of selection compared with other similar systems. Thus the drops which were preserved for further evolution would be just those which had a structure enabling them to adsorb porphyrins selectively.

Analogous considerations also apply to the formation of other specific catalysts and enzymes. Even such comparatively simple substances as, for example, methylamine, glycine, aldehydes, sugars, etc., have a weak catalytic activity for some reactions. These compounds could enter into the original systems or even, to some extent, be synthesised there. In the various coacervate drops they could combine with one another and with the inorganic catalysts present there in hundreds and thousands of different ways. Among all these combinations there must certainly have been some in which the catalytic activity was greatly increased owing to a favourable disposition of active and activating groups. A particular case might be the successful combination of amino acid residues in the polypeptide chains of the protein-like substances. This might give marked advantages to the systems in which there were formed combinations which had a powerful catalytic activity favourable to their dynamic stability and general activity.

This internal chemical rationalisation of the systems was reinforced by their selection. This destroyed those in which there had arisen, by chance, 'unsuccessful' combinations which diminished the catalytic activity. It preserved for further evolution only the more efficient catalysts which were more capable of performing their functions.

We have pointed out above that the very highly developed structures of catalase, carboxylase or any other enzyme could not have arisen by the action of selection on their separate isolated molecules, because the reactions which they carry out are of no significance to the catalase or carboxylase themselves. Their hastening or slowing of reactions cannot be reflected in the length of the existence or an increase in the amount of the enzymes as such. This activity may, however, have a decisive effect on the existence of the system in which any particular catalyst acts. Thus these systems must have been selected for the characteristic in question, and thus there could have arisen that extreme 'purposefulness' of structure, that correspondence between structure and function, by which enzymes may be recognised as biological formations.

Indeed, although we are now very rapidly approaching a full understanding of the chemical nature of enzymes, and even the solution of the problem of their synthesis by artificial means, these catalysts still bear all the marks of their biological origin. In nature they are only to be found in organisms and can only be formed naturally there. Such a 'fortunate' combination of atomic groups as we find in enzymes, such an intimate association between their structures and their biological functions, could not have arisen by chance or simply as a result of the action of the laws of physics and chemistry. The formation of enzymes required a definite orientation of the process of the evolution of matter, it required selection, the destruction of all 'unsuccessful' combinations and the retention for further evolution of only those systems in which the catalytic apparatus fulfilled its biological function most rationally.

This evolution of enzymes is still taking place to some extent. It must, however, be pointed out that the basic forms of construction of catalytic systems were already elaborated at what, comparatively speaking, was a very early stage in the establishment of life and its further development. Nowadays, therefore, even in the most poorly organised of contemporary living things, the individual enzymes are present as fairly highly-developed formations.

Origin of the Co-ordinated Networks of Reactions

Enzymes are, however, only the elementary and simplest form of organisation of protoplasm, its separate working mechanisms. The extreme specificity of protein enzymes means that each of them can only form intermediate compounds with a definite very narrow group of substances and can only catalyse strictly determinate individual reactions. However, the separate reactions catalysed by the different enzymes cannot of themselves, in isolation, serve as a basis for the process of life. Their biological significance becomes manifest and well defined only by virtue of their strict co-ordination with all the other chemical processes of the living body; their place in the general network of reactions in open systems is only maintained when they are included as essential links in a long chain of metabolic processes.

Hundreds and thousands of enzyme proteins play their parts in each vital process, let alone metabolism as a whole. Each can catalyse only one or a very limited number of reactions, and it is only when taken together, when their actions are unified in a definite way, that they constitute the orderly sequence of phenomena which forms the foundation for the process of life.

By using chemically individual enzymes isolated from living organisms one may produce, under laboratory conditions and in isolation, separate biochemical reactions which are links in the metabolic chain. This enables us to unravel the complicated skein of chemical reactions which make up metabolism, in which thousands of individual chemical reactions are carried out; to dismember metabolism into its constituent stages; to analyse not only the composition of living bodies, but also the chemical processes which are carried out in them and on which vital phenomena depend.

The great service of A. N. Bach (Bakh) to biochemistry was that, as early as the end of the nineteenth century, he showed, in his study of the chemistry of respiration, for example, that this phenomenon could not depend on the effect of any single enzyme (e.g. laccase or some other oxidase) but consisted of a chain of enzymic reactions which followed one after the other and were co-ordinated in an orderly fashion.

The same thing was established somewhat later for another important vital phenomenon, that of fermentation.

L. Pasteur in his day said that:

> *The chemical act of fermentation is essentially a phenomenon associated with a vital activity, beginning and ending with that activity; there is no fermentation without simultaneous organisation, development, multiplication of globules or the continuation of life by globules which are already formed.*

This supposition was refuted experimentally at the turn of the century by E. Buchner. By using a high pressure, he expressed a juice from yeast which did not contain any living cells but which could nevertheless ferment sugar. Buchner believed that his juice contained a specific enzyme, 'zymase', which broke the sugar down to alcohol and carbon dioxide by a single chemical act, just as, for example, invertase breaks sucrose down into glucose and fructose. However, the work which continued to be carried out for many years afterwards by a whole constellation of the outstanding biochemists of the first half of the present century, in particular by S. Kostychev, A. Lebedev, C. Neuberg and O. Meyerhof, showed that Buchner's juice contains, not one single enzyme, but a whole complex of such catalysts. Each of these accelerates its own specific reaction. All these reactions are combined together to form a long chain of transformations following one another successively in such a way that the end product of the preceding reaction serves as the starting substance for a rigidly determinate succeeding reaction. Sugar, on the one hand, and carbon dioxide and alcohol, on the other, are merely the first and last links of this chain. The reaction catalysed by each separate enzyme of the zymase complex occupies its own essential place in the chain of transformations, and forms an indispensable part of the chain as a whole. By poisoning or blocking, one may inactivate selectively any single enzyme of the zymase complex and thus exclude the reaction which it catalyses from the general sequence. The whole chain is then immediately disturbed and fermentation ceases or is distorted.

By now most of the enzymes of the zymase complex have been studied in great detail. Many of them have been isolated and obtained in a pure state; their chemical nature, the character of their specific action and their dependence on a number of physico-chemical conditions have been established. Alongside this analytical work a number of extremely interesting studies have been made, reproducing not merely individual enzymic reactions, isolated links in the chain of fermentation, but whole concatenations of these links, combinations of successive reactions catalysed by several enzymes of the zymase complex. Thus it seems to be possible to reproduce alcoholic fermentation artificially by the simultaneous action of all the enzymes and co-enzymes isolated from Buchner's juice.

During the interaction of our original colloidal systems with the medium surrounding them, and in the process of their later development, there must have been formed within them, not a single individual enzyme, but many specific catalysts. Their simultaneous activity determined the occurrence of some particular chain of chemical reactions or a whole network of reactions. On the nature of the organisation of this chain or network depended the greater or less dynamic stability conferred by the network on the open system. The selection of systems was based on this stability, destroying those which had an 'unsuccessful' combination of reactions and preserving for further evolution only systems with chains and networks which enabled them to survive for a long while under conditions of constant interaction with the external medium. It is obvious that it required a very prolonged and rigorous selection of a colossal variety of such systems for there to arise, at last, a chain consisting of more than 20 rationally concordant reactions such as take place in alcoholic fermentation. In principle, however, the origin of such a harmony between different catalytic reactions could quite well have occurred during the process of directed evolution and it seems that it must have come about at a comparatively early stage in the origin and development of life since the same basic collection of chains is common to, literally, all representatives of the living world which have been studied in this respect.

However, the form of organisation of the chain of processes on which extracellular fermentation is based is still relatively primitive. It is only based on a certain qualitative composition of the mixtures of enzymes, i.e. the obligatory presence in it of the whole collection of enzymes of the zymase complex. The sequence of reactions in extracellular fermentation simply depends on each intermediate product having its own specific enzyme. Other transformations of the product are excluded because, in the absence of the corresponding catalyst, they would proceed incomparably more slowly than the reaction which is accelerated by the enzyme. For this reason the whole process of extracellular fermentation is of the nature of a straight, unbranched chain. In the living cell it is of great importance not only what enzymes are present but also what are the quantitative relations between the various catalysts acting there; there must always be a certain correspondence between their activities. This is specially important when one and the same substrate can interact with several of the enzymes present in the cell. As a result of this the substrate is, in fact, altered in different directions. The chain of reactions then becomes branched, and the relationship between the rates at which reactions occur in the different branches has sometimes been found to determine whether or not some vital process can take place. A small change in this relationship may cause not merely the cessation of a process, but even the disruption of the whole system. It only takes place normally when the process of oxidation of the chromogens into respiratory pigments by the oxygen of the air and the reverse process of their reduction at the expense of the hydrogen of the appropriate donors, correspond very closely with one another, when their rates bear a precisely determined relationship to each other. If, as happens on mechanical injury to the cell, the rate of oxidation is increased disproportionately to that of reduction, the respiratory pigment will not be able to be reduced and will undergo further oxidation into a stable brown pigment which cannot serve as a hydrogen acceptor. In this way all the chromogen of the cell is very quickly converted into an inactive state and the process of respiration ceases as a result of the disturbance of the mechanism on which it is based.

There is a great variety of substances in protoplasm by means of which the accurate regulation of the catalytic activities of the enzyme complex is accomplished. In addition to the new formation and irreversible destruction of the protoplasmic enzymes, there also occurs a widespread reverse activation or inhibition of these catalysts. The protein nature of enzymes not only determines their exceptional activity and the specificity of their effects, it also determines their great lability, their extreme sensitivity to different kinds of physical and chemical factors. Any rough treatment will cause the denaturation of proteins and their catalytic activity will be irreversibly lost. But by treatment which does not lead to denaturation the activities of enzymes may be altered reversibly over a very wide range. In fact there is no physical or chemical factor, no organic compound or inorganic salt, which cannot affect the course of enzymic reactions in one way or another. Any raising or lowering of the temperature, any change in the acidity of the medium, its oxidationreduction potential, its salt content or its osmotic pressure, interferes with the relationship between the rates of the different enzymic reactions and thus changes their interconnections in the network of metabolism. Of great importance in this connection is the development among the components of protoplasm of various activators and inhibitors with specific activities, which selectively speed up or slow down any one or several enzymic reactions.

Owing to the action of all these supplementary chemical mechanisms which are intimately associated with the physicochemical state prevailing at any given moment within the protoplasm, very precise quantitative relationships are established between the rates of the enzymic reactions. These relationships may, however, vary greatly both as between different organisms and even in a single cell at different periods of its existence, and owing to the effects of different external and internal conditions. This gives a form of organisation which is very labile and adaptable, but at the same time very efficient. The process of extracellular fermentation is not associated with any protoplasmic structure; the whole process simply takes place in a solution of the enzymes of the zymase complex. In the cells of contemporary organisms, on the other hand, the spatial organisation of their protoplasm exercises a great, and sometimes decisive, influence on the rate and direction of the enzymic reactions on which its metabolism is based. We now know that the enzymes of cells are present, for the most part, in an associated state on protoplasmic surfaces and various cellular structures.

The investigations carried out in the Institute of Biochemistry of the Academy of Sciences of the U.S.S.R. have shown that the degree of association of the enzymes with the structures mentioned has a decisive effect not only in determining changes in the rates of the reactions catalysed by the enzymes, but also in displacing the dynamic equilibrium of the chemical processes towards a predominance of breakdown or synthesis. This, naturally, is of paramount importance for the self-preservation and growth of the whole living system. Phenomena of this sort cannot be explained on the basis of the laws which have been established for closed systems. As was shown in the previous chapter, however, in open systems (as distinct from enclosed ones) a catalyst may alter the stationary concentrations of the reacting substances, i.e. it may displace the experimentally determined, dynamic 'equilibrium' of the process.

This sort of influence of the protoplasmic structures on the rate and direction of the enzymic reactions of the metabolic network leads to a very intimate and critical connection between the metabolism and the conditions of the external medium. It very often happens that a factor which has a very weak or hardly noticeable effect on the activity of isolated enzymes will produce a radical displacement of the equilibrium between breakdown and synthesis, by altering the associated power of the protein structures of protoplasm, which are very sensitive in this respect.

According to contemporary cytological evidence a very considerable part of the cytoplasm, up to 50 per cent of its weight, is composed of different formed structures, particles of various sizes, for the most part mitochondria and microsomes. The mitochondria are rod-shaped formations visible under the microscope. Their internal structure has been fairly well studied both as to its morphology and its physical chemistry. They have an envelope consisting of two protein layers with a lipid layer between them. The internal core has a complicated structure and is also made up of proteins and lipids. The microsomes are submicroscopic and can only be discerned with the electron microscope. Their structure has still only been very poorly studied. According to J. D. Bernal the arrangement of the molecules of protein and nucleic acid in them is reminiscent of the structure of globular virus particles.

Both the mitochondria and the microsomes are very rich in lipids. The mitochondria contain the iron-porphyrin systems of the cell while the bulk of the nucleic acid is situated in the microsomes. The mitochondria contain large amounts of various enzymes. They seem to embody the catalytic mechanism required by the cell for the processes of oxidation and decomposition which lead to the liberation of energy from the multifarious substrates entering the cell, and also for the processes of transformation of this energy into forms in which it can be used in synthetic processes and for carrying out work in general. In particular, in the cells of highly developed organisms capable of respiration, this is carried out by means of the tricarboxylic cycle of Krebs. This is the most widespread system of oxidation of intermediate products of the breakdown of various organic substrates. The Krebs cycle comprises a strictly ordered concatenation of a large number of enzymic reactions, especially the hydration, dehydrogenation and decarboxylation of organic acids. At particular points in the cycle there branch off side reactions leading to the formation of substances which can serve as material for the synthetic processes of the cell. For example, α-oxoglutaric acid is one of the links in the Krebs cycle. It is formed from oxalosuccinic acid and later, in the course of the transformations of the cycle, it is converted, by oxidative decarboxylation, into succinic acid. This in its turn is transformed into fumaric and then into malic acid, etc. The α-oxoglutaric acid may, however, be transaminated and part of it may leave the cycle by a side route and be converted into glutamic acid which later serves as a material for the synthesis of proteins. In this way part of the α-oxoglutaric acid is always leaving the cycle irreversibly. In just the same way other keto acids (pyruvic and oxaloacetic acids) react with ammonia, i.e. are aminated directly, or are transaminated to form alanine and aspartic acid respectively. These are later transaminated to form other amino acids.

The acetic acid arising in the cycle may later take part in the formation of the citric acid of the cycle. Alternatively it may leave the cycle to serve as the starting material for the formation of fatty acids and other lipids. Oxaloacetic acid and glycine are essential materials for the biosynthesis of purine and pyrimidine bases, and glycine and succinic acid for the construction of porphyrins. Thus, all these biosynthetic processes which form the basis for the synthesis of living protoplasm are intimately associated with catabolism, from which they obtain their structural starting materials.

At certain definite points on the cycle there is also liberation of energy which is derived from high-energy bonds. The energy of the substances in which these bonds were originally present is transferred to ATP in the mitochondrial system. This, in its turn, activates substances taking a direct part in synthetic reactions.

Owing to the extremely efficient spatial disposition of the enzymes and coenzymes of the respiratory and energetic complex, their orderly assembly in the mitochondria, the cell achieves a maximal effect in the oxidation of substrates and the transformation of energy. The energy made available in this way is intimately associated with the formation of fragments of molecules which serve as materials for the synthesis of the substances of which the cell is made. The intensity of the oxidative processes in the mitochondria is, therefore, regulated by factors responsible for maintaining the balance between the liberation of energy and the supply of materials required for synthesis.

The direct synthesis of proteins from amino acids takes place in the nucleic acid-rich microsomes, as has been shown by experiments using labelled atoms. The energy needed for this reaction is made available in the mitochondria. It has been supposed that it enters the microsomes in the form of high-energy bonds of activated peptides containing the α-glutamyl group. The microsomes themselves seem to be formed from proteins, ribonucleic acids synthesised under the control of the nucleus and lipids which are formed in the mitochondria and stored in the Golgi apparatus.

These interactions of the formed elements of the cell is given by Lindberg and Ernster. The form of organisation of protoplasm which we have described is extremely efficient. With an organisation of this sort, owing to its definite spatial localisation, complete independent blocks of well 'assembled' enzymes interact in the performance of certain vital functions. Thus we have here a well-organised 'division of labour' between the various structures of the cell operating to achieve the maximum effect in the transformation of energy and the synthesis of living material.

Naturally this sort of organisation could only arise as a result of the prolonged development of living matter and it is, therefore, only to be found in organisms which have already reached a comparatively high level on the evolutionary ladder. Even in the comparatively poorly organised living things of the present time it is far more primitive. It must be supposed that the organisation of the primaeval organisms was even more primitive, although such a form of spatial localisation of the various enzymes and the reactions which they catalyse must have existed even at this stage of evolution.

Even the reactions of alcoholic fermentation take place far less harmoniously in Buchner's juice, where the spatial localisation of the enzymes is largely destroyed, than in yeasts or bacteria. But, most important of all, in Buchner's juice the process of fermentation follows, as it were, a lone trail. Here none of the energy liberated during the breakdown of sucrose to carbonic acid and alcohol is used rationally in any way. In the living cell, on the other hand, owing to the strict co-ordination of the chemical reactions, this energy participates to a greater or lesser degree in the process of synthesis of living material.

Of course, it is still very hard to answer the question as to what was the spatial organisation of the earliest living things. Considerable light might be shed on this problem by a comparative study of this organisation among the more primitive contemporary organisms. The study of multiple coacervates might also give some indication of the possible means whereby the simplest internal structure of the original colloid systems could have arisen. In multiple coacervates formed of several components there is an internal separation of the individual components in space, in that small droplets of one coacervate arise within the drops of another. For example, on mixing solutions of gelatin, gum arabic and sodium nucleate, drops are formed, composed of gelatin and gum arabic. Within these drops there are formed small droplets containing gelatin and nucleic acid. This can easily be demonstrated by selective staining or by the use of the ultraviolet microscope. Various substances and catalysts may become localised on the internal surfaces which are formed in this way.

In summarising what has been said one must emphasise the extreme complexity and diversity of the factors which determine the organisation of contemporary living bodies in time, the causes on which the structure of the network of chemical transformations of their metabolism depends. The formation of this network was determined by the chemical properties of the compounds of which living bodies were composed. The great diversity of these compounds and their extreme chemical reactivity carried with them the possibility of numerous chemical transformations and unlimited combinations of the compounds. But, in this extremely wide field of chemical possibilities the process of directed evolution, by the gradually increasing organisation of living systems, led to the emergence of ever more clearly defined pathways of biochemical processes which formed a more and more efficient network of metabolic reactions.

In contemporary organisms this network has reached a very high efficiency. Its organisation is determined, as we have seen, by a whole complex of concordant factors: the presence of a particular collection of enzymes, their quantitative relationships, the physico-chemical conditions prevailing in the protoplasm, its colloidal properties and, finally, its structure, the definite localisation of chemically and biologically active compounds and the irreversible nature of the biochemical processes. The original systems, and even the earliest living things, did not have to the full such a complicated and efficient form of organisation. However, both before and after the emergence of life, there took place a directed evolution, not of isolated factors or parts of the system, but of the metabolic network as a whole, leading towards its improvement. In the course of this evolution there continually arose new pathways, some of which became predominant in metabolism at the same time as old pathways disappeared or merely remained in reserve. During all these changes, however, there was always maintained a network which, to some extent, provided for constant self-preservation and self-reproduction of the system as a whole. The improvement of the metabolic network only implied the more and more rational performance of this task under more and more diverse and varying environmental conditions.

As the metabolic network improved so there arose and developed those properties of the system which may be regarded as the characteristic features of life, which are fundamental to the organisation of this form of the motion of matter.

The interaction with the external medium of such systems as coacervate droplets, which have no organised network of chemical reactions, can only be based on the permeability of surface membranes or the adsorptive properties of colloids. In this case, however, the entry of substances into the system by such means soon ceases and the system enters into equilibrium. Only when the substances entering the system in one way or another can be changed and accumulated within the system in the form of some particular compounds, or cast out into the surrounding medium as breakdown products, can the phenomenon of interaction between the system and its environment continue for a long time. This must have occurred when coacervate drops were transformed into open systems at a comparatively early stage in the evolution of the original colloidal formations. But, in this case, the entry of substances into the system or their expulsion into the external medium must already have ceased to depend on the simple laws of permeability and adsorption and have depended on the state of development of the organisation of the network of reactions into which the substances derived from the external medium entered, or in which the breakdown products which were expelled were formed.

It is precisely this sort of interaction with the external medium, though in a considerably more highly developed form, which is characteristic of all contemporary living things. According to the evidence of contemporary cytology and cellular physiology, the entry of substances from the environment into the cell is not a passive process determined by the greater or lesser mobility of these substances through a hypothetical semi-permeable membrane, or by their selective adsorption on protoplasmic surfaces, as had earlier been supposed. This entry is brought about by the active participation of the whole cellular metabolism. It occurs because the substances which enter are drawn into the network of metabolic reactions. For this reason any disturbance of metabolism, such as a decrease in cellular respiration, has an immediate and decisive effect on the entry of substances into the cell.

In an analogous way the characteristic features of cellular energetics depend on the high degree of organisation of the metabolism of protoplasm. In the engines which are widely used in industrial processes, the chemical energy which is liberated by burning the fuel is usually first converted into heat and only later transformed into other forms of energy. In protoplasm the energy liberated by the decomposition of organic substances (in the process of fermentation or by their oxidation during respiration) is converted directly into the forms of energy required for life. Owing to this an extremely high coefficient of utilisation of energy is achieved in living bodies such as is not approached by our technology. In the engines of the present time this coefficient reaches, at best, 40 per cent, and this requires considerable temperature differences, of the order of hundreds of degrees. If the transformation took place in living bodies in the same way as in heat engines, then, at the temperature differences which are possible for organisms, the coefficient of energy utilisation would only be a fraction of 1 per cent. Nevertheless it in fact reaches 5 per cent or even more. This is explained by the fact that the breakdown and oxidation of sugar or other energy-yielding material does not take place as an isolated process in the living cell, but through a series of separate reactions which are strictly co-ordinated in time and which form the chains and cycles which constitute metabolism. The chain of alcoholic fermentation and the oxidative cycle of Krebs may serve as examples.

It must be pointed out that if the oxidation of organic molecules were to take place all at once in protoplasm, the living body would not be able to make rational use of the energy thus liberated. The oxidation of only one gram-molecule of sugar to carbonic acid and water liberates about 700 kcal. The instantaneous release of this amount of energy would be associated with a sharp rise of temperature, the denaturation of proteins and the destruction of protoplasm. The energetic effect achieved by protoplasm at ordinary low temperatures depends on the fact that, in the process of biological oxidation or degradation of sugar, this substance is not converted into its end products all at once, but by gradual stages.

This sort of organisation not only gives rise to the possibility of overcoming the high barrier of the energy of activation of the reaction of the oxidation of sugar by atmospheric oxygen at ordinary temperatures; it also allows the living cell to make rational use of the energy, which is not liberated all at once but gradually in separate small portions. We have already seen from the example of the Krebs cycle that this is what actually takes place. At definite points on this cycle energy is liberated, whereupon it is immediately taken up to form ATP, or some other compound with high-energy bonds, which may be used for carrying out syntheses or for performing work necessary for life.

The more highly organised the metabolism and the better the co-ordination between the separate reactions of which it is made up, the higher will be the coefficient of useful work. Direct observations on various representatives of the living world show that in poorly organised living things, standing at the bottom of the evolutionary scale, the reactions of the energetic network are not always strictly co-ordinated. A considerable amount of the energy liberated in them is therefore dispersed aimlessly and cannot be used for vital processes, in particular for the formation of new living material and the growth of cells. When, on the contrary, the rates of the reactions are strictly co-ordinated, when they are, so to speak, accurately adjusted to one another, this waste of energy is cut down substantially. In such a case a relatively small access of the organic materials serving as the source of nourishment leads to considerable growth of the living thing.

This may be seen in moulds, for example, where the metabolism is very highly organised. H. Tamiya in particular obtained the following data for Aspergillus oryzae: for the formation of 1 g of mycelium the mould assimilated 1.467 g of glucose. The efficiency of the utilisation of energy was thus in moulds. In some bacteria, on the other hand, this efficiency is incomparably less and this is associated with incomplete co-ordination of the various links of their chains of energy metabolism.

The energy metabolism of the primaeval organisms must certainly have been at a still lower level of development. The commencement of this metabolism must, however, have taken place in the very early stages of the evolution of our original systems as the only easily mobilised sources of energy available to them were organic compounds which entered the system from outside. However, if the energy contained in these substances was to be released, they had to be broken down in some way. At first this breaking down took many forms and followed various paths and its efficiency must therefore have been very low.

Later, however, selection led to the formation of several standard paths which formed the basis of the energy exchange of all living things without exception. The earliest reactions in the process of glycolysis, in particular, seem to constitute such paths. They have been found in all organisms where they have been looked for.

As we saw above, the definite organisation of the network of metabolic reactions also forms the basis for the synthesis of all the substances formed in living matter. This may be demonstrated above all by the origin of the property which first Pasteur and then Vernadskii considered to be one of the most characteristic features of life, namely, the asymmetry of protoplasm.

What is the cause of this asymmetry? Why is there only formed in living protoplasm one particular optical configuration of amino acids and other similar compounds?

We indicated that in the original solution of organic substances the action of circularly polarised light or selective synthesis on the surfaces of quartz crystals could have led to the appearance of a certain dissymmetry, some predominance of the *dextro* or *laevo* antipodes of particular compounds. This original asymmetry may have formed the basis for the asymmetry of all later organic formations.

It is generally supposed that one molecule with a particular optical configuration gave rise to another exactly similar molecule, 'multiplied' so to speak more and more, so that there was a steady increase on the surface of the Earth in the amount of the compounds which belonged to the series of optical isomers in question. A detailed examination of the subject shows, however, that the matter is considerably more complicated.

W. Kuhn in his day undertook a detailed analysis of all the evidence then available concerning asymmetric synthesis. He showed, in the first place, that a racemic mixture is thermodynamically more stable than its separate optically active components because the free energy is less in the racemic mixture. Any mixture of optically active substances will, therefore, tend to racemise and lose its optical activity. In any synthesis mediated by an asymmetric catalyst (e.g. an enzyme) at first only one optical antipode will be formed quickly. However, the other antipode is formed too, but in an amount which is as many times smaller than that of the first as the rate of the synthesis in the presence of the catalyst is greater than the rate without the catalyst. This will lead to the appearance of a certain asymmetry, a certain inequality between the amounts of the *dextro* and *laevo* antipodes. But as a true catalyst increases the rate of a reaction and of the reverse reaction, after equilibrium has been reached some of the product of the synthesis will be converted into the starting substance giving rise, though very slowly, to fresh amounts of the antipode which is synthesised without the catalyst. Thus the whole system will tend towards the racemic state and the optical activity which arose as a result of the action of the asymmetric catalyst will gradually get weaker as may, in fact, be demonstrated by experiment.

Thus the asymmetry which arose in an enclosed system owing to the activity of any isolated reaction must have been a temporary phenomenon and could not have served as a basis for the formation of the very complete and constant asymmetry of protoplasm.

The continual formation of only one optical antipode can only occur in open systems on the basis of a definitely organised network of reactions, the rates of which are very accurately related to one another. Under such conditions, when there is a definite sequence of processes, racemisation or the appearance of the opposite antipode may be avoided altogether and the system may retain its asymmetry indefinitely.

Thus the asymmetry of protoplasm is due to its definite organisation in time, the co-ordination of the reactions occurring in it. This co-ordination certainly did not arise by chance. It was enabled to arise by selection of the original systems, for, according to W. H. Mills, everything else being equal, reactions proceed at a significantly slower rate in racemic mixtures than in optically active mixtures. Systems made up of asymmetric material must therefore have been more effective in the struggle for existence than their competitors made up of racemic mixtures.

Systems, in which a definite co-ordination of reactions had led to the formation of asymmetry, carried out their syntheses more quickly than otherwise identical systems based on racemic mixtures. The growth of the former must therefore have been significantly faster and their dynamic stability must have been greater. As a result of all this the action of selection must have tended to increase the asymmetry of the substances entering into the composition of our original colloidal systems from the first stages of their evolution. In contemporary protoplasm this asymmetry has reached a very high level; there is an extremely high degree of optical purity which can only be present as a result of a very close coordination of the rates of the reactions contributing to the synthesis of the substances in question.

In order to renew and preserve themselves continually in a state of uninterrupted interaction with the external medium, our original systems could extract the ingredients which they needed ready-made from the medium, sometimes in the form of very complicated organic molecules. However, even at the earliest stages of the evolution of our original systems, the substances entering them must have undergone some sort of chemical transformation, otherwise the properties which characterise open systems could not have arisen. As we saw above, this did not involve any direct 'multiplication' of the individual molecules, they could not 'reproduce themselves' directly. What did occur was only a more or less constantly repeated formation of new material based on chemical reactions taking place in a certain sequence. Thus, for the progressive evolution of our original systems the important thing was not the chance entry or development of any particular compound, but the appearance of a definite co-ordination of the reactions which provide the constant synthesis of this compound in the system in continuous interaction with the external medium.

The more closely the chemical substances entering it from the external medium resembled ingredients of the system itself, the less complicated were the chains of reactions leading to its synthesis and the simpler was the metabolic organisation. However, the system was correspondingly more dependent on the constancy of the external medium and on its high content of complicated organic compounds.

Clearly the selection of our original systems and the emergence of the first organisms from among them must have been directed towards a lessening of this dependence and the formation of networks of synthetic reactions by which the complicated ingredients of the system could be synthesised unerringly from the somewhat diverse compounds which entered it from the external medium. For this purpose the compounds in question had to be 'standardised', that is to say, broken down to relatively simple and uniform fragments from which any specific ingredients of the system could be built up by standard methods, though using complicated chains of transformations with many links.

We do, in fact, find such a form of organisation in the constructive metabolism of contemporary organisms. In them, as we have pointed out several times, very simple compounds of low molecular weight such as oxalic acid, glycine, succinic acid, keto acids, etc., serve as starting points for the synthesis of proteins, nucleic acids, lipids, porphyrins and the other complicated ingredients of protoplasm. These simple compounds arise as fragments split off in the course of the destructive metabolism of the sugars and other substances which enter the cell from its surrounding medium and serve as its nutrients. In the course of this destructive metabolism the energy needed for synthesis is liberated and stored in ATP and other compounds with high-energy bonds. We have already explained this with reference to glycolysis and the tricarboxylic acid cycle of Krebs. These examples show that the biosynthetic processes, which form the basis for the formation of the living material of organisms, are intimately associated with destructive metabolism, from which they obtain their original structural materials and the energy needed for synthesis.

The route by which the fragments under discussion are built up into proteins, nucleic acids or porphyrins is, of course, very complicated and consists of a series of successive reactions. Such syntheses can, therefore, only take place by means of a very precise and absolutely constant co-ordination of these successive reactions, by means of a very highly developed organisation of the metabolic network. This was also essential for the appearance of the most characteristic feature of life, the capacity for self-reproduction.

Thus, we can now already give an indication, though still only a very rough and speculative one, of the actual course of development leading from the initial systems to the simplest organisms during the emergence of life on our planet. This development involved successive improvements in the networks of reactions within individual colloidal systems which were reacting with the external environment. Owing to continual changes in these systems, within the limits of their dynamic stability, they underwent the following transformations. First there was the formation of individual catalysts of great reactivity and specificity. Later the activity of these catalysts was co-ordinated and there arose the whole chains and cycles of enzymic reactions which form the basis for the separate departments of metabolism. Still later came the spatial organisation of the system and the localisation of processes and the rationalisation of the interacting energic and structural branches of metabolism. This guaranteed, within limits, the continual self-preservation and selfreproduction of living systems.

The nature of this organisation may, naturally, vary within limits in different representatives of the living world, but it is always an expression of the degree of integration attained by the organism in the course of its evolutionary development. A comparative study of contemporary living things will enable us to form an opinion as to the course of the successive integration of this remarkable form of the motion of matter which came into being at some time on the Earth.

4

Theory of Eternity of Life

Introduction

The principle of the eternity of life is not incompatible with the possibility of spontaneous generation of living creatures. As we have seen in the previous chapter, idealists have, from ancient times, united the two doctrines. This union was specially clearly expressed in the doctrine of '*panspermia*'. According to this, the fertilising or life-giving principle takes the form of invisible spiritual germs of life dispersed everywhere.

We first encounter the actual term 'panspermia' in the work of the ancient Greek philosopher Anaxagoras (500-428 B.C.). In his view, the various living creatures originate from slimy earth when it has been fertilised with 'ethereal germs' (*spermata*) which are present everywhere. Later on, the doctrine of panspermia acquired a markedly idealistic character. We find it in this form in the teachings of Roman philosophers, of 'the fathers of the Christian Church', of the mediaeval schoolmen and of a number of more recent natural philosophers.

The works of St. Augustine of Hippo may be taken as an example. He held that the earth is full of hidden life engendering forces, *occulta semina*, invisible, mysterious seeds of spiritual origin, which become active under favourable circumstances and produce plants, frogs, birds and insects from water, air and earth. The 'spirit of growth', *anima vegetativa* of the later scholasts, the arche of Paracelsus and van Helmont, the 'life force' of a number of other authors, etc., were also of this nature.

In the middle of the seventeenth century Athanasius Kircher developed his theory of panspermia, according to which the germs of life are scattered in chaos and in all the elements, and the various animals and plants arise as a result of their activity. A principle similar to that of panspermia forms the foundation for Leibnitz' teaching concerning the immortal, ubiquitous germs of life which, in the course of their later development, form all living things. According to Needham, the vivifying principle 'life force' is inherent in every particle of organic matter and only under its formative influence can micro-organisms develop in decaying materials.

Pouchet took up an analogous position. He considered that spontaneous generation was only possible as a result of the action of the 'life force' which had previously entered the molecules of organic substances. When the theory of spontaneous generation was exploded towards the end of the nineteenth century the vitalists and neovitalists quietly abandoned it, bringing to the fore the principle of the eternity of life and emphasising the impossibility that the human mind could ever solve the problem of its origin.

The position was different for those natural philosophers who were working on a materialistic basis. They were trying to use the theory of the eternity of life as a way out from what seemed to be the *impasse* which had been created by Pasteur's experiments. It is clear that the theory of the eternity of life as something which has a separate existence, divorced from matter, is foreign and hostile to materialism. Mechanistic materialism and, in particular, hylozoism, assume the eternity of life, and regard it as merely a constant and inalienable property of matter in general. If we accept this, the spontaneous generation of living creatures follows *ex hypothesi*. If all matter is endowed with life, if there is, in principle, no qualitative difference between organisms and objects that are inorganic in nature, then living creatures must inevitably arise spontaneously, even in the absence of other living creatures. Hylozoism without spontaneous generation is absurd. It is thus inconsistent for materialists to make use of the theory of the eternity of life to explain the impossibility of spontaneous generation. This leads inevitably to idealism.

Clear examples of this attitude are found in the pronouncements of a number of authoritative scientists of the late nineteenth and early twentieth centuries. Many of these scientists regarded the experiments of Pasteur as proof of the absolute impossibility of the metamorphosis of inorganic materials into living organisms. In 1871 the distinguished British physicist W. Thomson, later Lord Kelvin, wrote in this connection: "Dead matter cannot become living without coming under the influence of matter previously alive. This seems to be as sure a teaching of science as the law of gravitation." Hence followed the complete autonomy of living creatures, and consequently also life must be regarded as eternal.

The famous German physiologist H. Helmholtz said: "It appears to me to be a fully correct procedure, if all our efforts fail to cause the production of organisms from nonliving matter, to raise the question whether life has ever arisen, whether it is not just as old as matter."

The French botanist van Tieghem wrote in his textbook: "The vegetation of the earth had a beginning and will have an end, but the vegetation of the universe, like the universe itself, is eternal".

We meet similar opinions among a number of other scientists who, proceeding from the empirically established fact of the impossibility of spontaneous generation, proclaimed that life is in principle eternal while still reckoning that they had based their position on materialistic principles. Thus, for example, the very able Russian plant physiologist and biochemist S. Kostychev wrote in the conclusion of his book *On the appearance of life on the Earth*: "When the echoes of the battle about spontaneous generation finally die away, everyone will recognise that life only changes its form, but never arises from dead matter". However, wishing to escape from the justifiable accusation of idealism, he added: "It must be noted that this point of view has nothing in common with the theory of vitalism, which is nebulous and hostile to progress". All the same, this denial is unconvincing, and it is not easy to see how one can combine acceptance of the eternity of life with denial of 'the eternal vital principle' or 'life force'.

As early as the late nineteenth century, F. Engels gave detailed consideration to the principle of the eternity of life, and showed convincingly that it is incompatible with consistent materialism. He quotes a very characteristic remark made by Liebig to M. Wagner in 1868.

We may only assume that life is just as old and just as eternal as matter itself, and the whole controversial point about the origin of life seems to me to be disposed of by this simple assumption. In point of fact, why should not organic life be thought of as present from the very beginning just as much as carbon and *its compounds* or as the whole of uncrea

and indestructible matter in general, and the forces that are eternally bound up with the motion of matter in space.

Engels points out that such views can only be based on recognition of a specific vital force, such as a 'formative principle', and do not at all correspond with a materialist picture of the universe. Engels further wrote in comment:

> *Liebig's assertion that carbon compounds are just as eternal as carbon itself, is doubtful, if not false. The compounds of carbon are eternal in the sense that under the same conditions of mixture, temperature, pressure, electric potential, etc., they are always reproduced. But that, for instance, only the simplest carbon compounds, CO_2 or CH_4 should be eternal in the sense that they exist at all times and more or less in all places, and not rather that they are continually produced anew and pass out of existence again in fact out of the elements and into the elements has hitherto not been asserted. If living protein is eternal in the same sense as other carbon compounds, then it must not only continually be dissolved into its elements, as is well known to happen, but it must also continually be produced anew from the elements and without the collaboration of previously existing protein and that is the exact opposite of the result at which Liebig arrives.*

The proposition that living beings invariably arise when certain conditions are fulfilled has nothing in common with the concept of the 'eternity of life'. On the contrary, it leads to the idea that organisms invariably originate from inanimate matter.

Against this, those who favour the eternity of life consider that at all times there has existed some element which has been passed in succession from organism to organism. Without this the occurrence of living beings is impossible. "Life," wrote F. J. Cohn (1828-1898), "is like the holy fire of Vesta, which was only kept in being continuously by kindling the new flame from the old." But what is this special principle that is present only in the living organism, and what is its nature? It cannot be an eternal property of matter, as the ancient Greeks supposed, because then the vivification of matter would not require the participation of a living organism already in existence, but life would arise spontaneously of itself. It cannot be a new quality arising in the course of the historical development of matter, because then it would not be eternal. Consequently, this principle cannot be material in nature. And so, as soon as we try to extend or develop the principle of the eternity of life, whether we want it or not, we find we have been trapped into idealistic assumptions. It cannot be said that attempts to resolve this contradiction on the basis of a so-called 'materialistic dualism' have been successful. This recognises the parallel and independent existence of two completely autonomous forms of matter, radically distinct from one another and separated by an impassable gulf.

The well-known Russian geochemist V. Vernadskii (1863-1945) presents the clearest example of this tendency. In his works written in the twenties and thirties of this century he puts forward the view that the idea "that logic demands that there should be a beginning of life came into science as a problem of religion and philosophy" and that it is "foreign to the empirical foundations of science". He wrote:

> *None of the exact relationships between facts which we know will be changed if this problem has a negative solution, that is, if we admit that life always existed and had no beginning, that living organisms never arose at any time or place from inert material, that in the history of the earth there were no geological periods in which life did not exist.*

Vernadskii held that the essential feature of the material and energetic characteristics of living bodies which distinguishes them from inert matter is that a special orientation is inherent in the former. He pointed out that even Pasteur recognised the possibility of different states of cosmic extension and that he used this concept to explain the phenomenon of asymmetry in living things, or, to use the terminology of *Vernadskii*, 'rightness and leftness'. This orientation which is associated with individual organisms is described by *Vernadskii* as follows: The mirror-image forms of each chemical compound are acknowledged to be chemically identical in inert matter and different in living organisms.

The chemical dissimilarity is thus conspicuous in the products of biochemical processes, in which either the *dextro or laevo* isomer predominates. *Vernadskii* further puts forward the idea that this orientation in space, which is associated with the body of the living organism, is only created in the biosphere from natural living bodies which have existed previously, that is, as a result of reproduction. Thus our lack of success in bringing about the synthesis of a living thing is due to the fact that the special asymmetric spatial conditions required for the purpose are absent from our laboratories.

The question of the 'rightness and leftness' of living substance deserves serious consideration and we shall return to it later, but it must be pointed out here that at present a large number of facts are being reported in the scientific literature which suggest the possibility of the production of asymmetric substances independently of living things in the presence of asymmetric factors acting in inorganic nature.

In one of his later works, published in 1944, *Vernadskii* seems to have taken account of these discoveries and did not refer to this difference between living and inert matter but only emphasised the fact that they differ in isotopic composition. The fact is that as early as 1926 *Vernadskii* demonstrated that the isotopic composition of the elements present in living organisms differs considerably from that of the elements derived from minerals and rocks. However, biogenic formations which arise in association with living things or after their death, such as soils, the waters of seas, rivers and lakes, petroleums, coals and bitumens, retain the isotopic composition characteristic of living things. *Vernadskii* therefore held that in this case one cannot *a priori* deny the possibility of transition of matter from the dead ('bio-inert') to the living state, 'for the atomic composition of the living and the inert matter may here be isotopically identical'. On the other hand, the direct transition from materials which have not arisen biogenically to living things would seem to be excluded on account of the profound differences in isotopic composition. However, as these biogenic formations ('bio-inert' substances) only develop in the presence of organisms a closed circle of life is set up.

One might infer from this that *Vernadskii* continued to believe in the complete impassability of the gulf separating the living from the lifeless, the complete impossibility of the primary origin of life from inert matter. However, such a conclusion would be premature. In the work which we have cited, *Vernadskii* shows convincingly, in a number of concrete examples, that a quantitative change in the isotopic composition of the elements "is not only characteristic of living matter but also occurs in processes which have nothing to do with life, as among the products of volcanic eruptions". The whole difference lies in the fact that changes in the isotopic composition of the elements brought about by organisms proceed on the surface of the earth at ordinary temperatures and pressures, whereas analogous changes in a lifeless medium only happen at high pressures and temperatures in the depths of metamorphic formations. "The synthesis of life", *Vernadskii* continued, "requires preliminary isotopic modification of the chemical elements". However, as we have just seen, *Vernadskii* himself pointed out that changes of this sort may occur in ordinary inert media at high temperatures and pressures and it is therefore quite arguable that life first originated from ordinary inert (not biogenic) matter under conditions where it was subjected to preliminary isotopic modification by the forces of inorganic nature.

Thus we have seen that, as a result of prolonged and varied studies of the question, *Vernadskii* abandoned the untenable position of 'materialistic dualism' which he previously held. In 1944 he wrote, "In our time the problem can hardly be treated as simply as it could be during last century when, it seemed, the problem of spontaneous generation had been finally solved in a negative sense by the work of Louis Pasteur."

It is hardly necessary nowadays to demonstrate theoretically the complete incompatibility of all kinds of dualistic views with a consistent materialism. We should, however, analyse in detail the factual evidence which has been and still is adduced in support of their attitude by the adherents of the theory of the eternity of life. We should examine how far this evidence agrees with the objective data of contemporary science. The chief difficulty which is always encountered by the materialistically inclined proponents of the eternity of life is the problem of the emergence of life on the Earth and of all those beings which inhabit the Earth. The Earth itself does not seem to be eternal, it originated at some time and it is therefore necessary to explain in some way how the first organisms appeared on it without recourse to the creative act of deity or the formative influence of a 'life force'.

For vegetation to develop on the virgin rocks of volcanic islands the seeds or spores of plants must have been carried there from elsewhere. A similar idea that viable germs from other worlds inhabited by organisms were deposited on the virgin earth during its development was put forward by the supporters of the theory under discussion as being the only possible explanation of the appearance of life on our planet. But before this hypothesis is scientifically admissible it must be shown that life is widely distributed throughout the universe, that it is to be found, not only on the Earth or within the solar system, but also in other parts of the universe. Furthermore, it is necessary to explain how the germs of life could be transferred to the Earth through interplanetary and interstellar space while remaining alive and able, under favourable circumstances, to grow and give rise to a new race of living things.

The bold suggestion that there might be a multiplicity of worlds inhabited by living creatures was very clearly stated by the great sixteenth century scientist Giordano Bruno. In his treatise *Del' infinito universo e mondi* he wrote, "There exist innumerable suns and innumerable earths circling round their suns just as our seven planets circle round our Sun. Living things dwell on these worlds." For a long time this idea did not spread far because it came up against the ancient but very active anthropocentric conviction that there is only one earth supporting life in the universe. It was considered daring and fantastic for a scientist to think that there might be many inhabited worlds. It is only 15-20 years since the authoritative English astronomer Sir James Jeans stated that,

> *We know of no type of astronomical body in which the conditions can be favourable to life except planets like our own revolving round a sun. Yet exact mathematical analysis shows that planets cannot be born except when two stars pass within about three diameters of one another. The calculation shows that even after a star has lived its life of millions and millions of years the chance is still about a hundred thousand to one against its being a sun surrounded by planets. All this suggests that only an infinitesimally small corner of the universe can be in the least suited to form an abode of life.*

Now, however, we cannot accept Jeans' point of view. On the contrary, contemporary scientific findings definitely confirm the inspired foresight of Bruno. In 1938 the Swedish astronomer E. Holmberg made careful analyses of a number of measurements of the right ascensions of stars the parallax of which had been determined with special accuracy. He demonstrated very small but definite oscillations with periods ranging from one and a half to three years. These oscillations can only be explained as disturbances caused by satellites of comparatively small mass. It would certainly be impossible to observe these satellites directly by means of present-day telescopes, but there is now no doubt that there are many stars which, like our Sun, are surrounded by circulating planets. Twenty-five per cent of the 240 stars observed by Holmberg give indications of the presence of small, invisible planets. Dark satellites having masses comparable with those of our own planets have already been discovered for many stars, e.g. 70 *Ophiuchi* and 61 *Cygni*.

It seems, therefore, that our solar system is not unique. There can be no doubt that planets revolve round other stars too, and very many of these are comparable with our Earth. There is therefore nothing to hinder us from supposing that life exists on some of them, maybe even on many of them.

In his book *Life on other worlds* H. Spencer Jones analyses a great deal of factual material relating to our problem and arrives at the conclusion that life is distributed throughout the universe and that the number of worlds where life is possible seems to be very considerable (see also the recent book of A. Oparin and V. Fesenkov, *Zhizn' vo vselennoi*. Moscow. Thus the first condition mentioned above for the acceptance of the theory under discussion, that is to say the wide dispersal of life in the universe, is not ruled out by the findings of contemporary science. The case is, however, different as regards the passage of the germs of life through space.

The hypotheses concerning this problem may be divided into two groups, (1) the transport of the germs by meteorites ('cosmozoë' or ' lithopanspermia') and (2) transport of the germs with cosmic dust under the pressure of light ('radiopanspermia').

The Theory of Çosmozë

The idea that fragments of stars bearing the seeds of life might reach the Earth and thus impregnate it was discussed as far back as the beginning of last century by the Frenchman de Montlivault. It was later developed by H. Richter in 1865. He started from the hypothesis that when celestial bodies are in rapid motion small pieces or solid particles may become separated or torn off from them.

It might be that the viable germs of micro-organisms were attached to the particles at the time when they became separated from the celestial bodies. Furthermore, these particles would wander in interstellar space and might, by chance, arrive on other heavenly bodies. When these germs fell on a planet where the conditions were favourable for life (suitable conditions of moisture and temperature) they would start to develop and, in the course of time, they would establish themselves as the ancestors of all living things on that particular planet. Richter assumed that somewhere in space there are always celestial bodies on which life exists in the form of cells. This idea was later developed by M. Wagner, who considered that "the atmospheres of the heavenly bodies, and also the swirling cosmic mists may be regarded as eternal repositories of living forms, as perpetual plantations of organic germs". Thus life is scattered throughout the universe and travels in the form of germs within meteorites.

Richter paid special attention to the possibility that viable germs might be carried through interstellar space. He pointed out that the germs of living things can exist for long periods without nutrients and water, remaining in a more or less inanimate state, and may then reawaken to a new life, though only when the necessary conditions are fulfilled. As a result of this capacity they may make very long journeys. The only hazard to which the germs of life are submitted arises from the increase in temperature which occurs as a result of the tremendous friction generated between the meteorites and the atmosphere of the Earth. However, Richter points out that some meteorites contain traces of carbon and other easily combustible substances. If these substances can reach the Earth without being burnt, it is perfectly possible that germs might pass through the atmosphere without losing their viability. Similar views were put forward in Britain by Lord Kelvin, who wrote in 1871:

> *Should the time when this Earth comes into collision with another body, comparable in dimensions to itself, be when it is still clothed, as at present, with vegetation, many great and small fragments carrying seed and living plants and animals would undoubtedly be scattered through space. Hence and because we all confidently believe that there are at present, and have been from time immemorial, many worlds of life besides our own, we must regard it as probable in the highest degree that there are countless seed-bearing meteoric stones moving about through space.*

These statements made a very, great impression on the scientists of those times. In Germany they were supported by H. Helmholtz, who considered that the germs of life had reached the Earth by means of meteorites which, in their passage through the atmosphere of the Earth, had been strongly heated on the surface only, while the inner part remained cool. In France this opinion was shared by van Tieghem, who wrote that the Earth received the seeds of life by their being carried on meteorites; henceforth it conserved the life which was derived from these original germs.

The main foundation for all these hypotheses was the fact that many rocky meteorites contain compounds of carbon approaching hydrocarbons in their composition. For example, chemical analyses by Cloez of the Orgeuil meteorite revealed the presence of amorphous substances very similar to the humus-like substances found in some fuels dug from the earth. At the time when the presence of hydrocarbons in meteorites was first discovered people were still convinced that organic substances, including hydrocarbons, could only be formed under natural conditions in living cells. Many scientists therefore supposed that the hydrocarbons found in the meteorites had been formed there secondarily as the result of the decomposition of organisms which had lived at sometime on these heavenly bodies. This raised the question of the possible existence of living bacteria or their spores inside the meteorites.

Nowadays, since the comprehensive investigations of D. Mendeleev and other chemists, we know that hydrocarbons and their derivatives can easily develop inorganically under natural conditions, particularly from cohenites, which are minerals commonly found in meteorites and composed of carbides of iron, nickel and cobalt (Fe, Ni, Co).

J. L. Smith showed that the organic substances found in the Orgeuil and other meteorites could have been formed as the result of reactions between iron carbide and iron sulphide. From the Orgueil meteorite Smith even prepared compounds of carbon, hydrogen and sulphur such as $C_4H_6S_5$. He showed that there is no foundation for the belief that these organic compounds have been formed by organisms.

Berthelot and Schutzenberger independently reached similar conclusions. They demonstrated in meteorites the presence of hydrocarbons completely analogous to those formed during the smelting of iron at temperatures which are certainly incompatible with life. Thus the discovery of compounds of carbon in meteorites cannot now serve as an argument that there are traces of life on these bodies.

Neither have numerous attempts to discover directly the germs of microbes on meteorites given definite positive results. S. Meunier stated that Pasteur, whom he supplied with specimens of carbon-containing meteorites, also tried to isolate viable bacteria from them. He even constructed a special boring apparatus for the purpose, which enabled him to take specimens from the inner parts of the meteorites. However, Pasteur always got negative results and therefore did not publish them. Later scientists have had no more success in finding living things in meteorites.

This communication attracted much attention in scientific circles and even found its way into some textbooks, but unfortunately it has not been confirmed up till now. It is worthy of note that the microbes obtained by Lipman seemed to be identical with the ordinary terrestrial bacteria. In view of the great variability of bacteria and the readiness with which they adapt themselves to external conditions, it is hard to believe that exactly the same forms of microorganisms exist on other heavenly bodies as on our planet. It seems far more probable that, in spite of all his precautions, Lipman failed to prevent terrestrial bacteria from falling on to the meteorites he was studying while he was grinding them. In a letter which he sent to me, Lipman himself did not insist that his results were completely unequivocal.

In the present state of our knowledge it is, in fact, hard to suppose that organisms are present inside meteorites. If life had developed at some time and place on the planet from which the meteorite had become separated, it would undoubtedly have left traces in the shape of biogenic formations. However, even after the most careful searches nobody has been able to find traces of such formations anywhere in meteorites. According to A. Fersman, F. Levinson-Lessing and others there is nothing resembling a sedimentary formation nor anything which might, in general, be ascribed to biological processes. Mineralogical studies of meteorites also show that they were formed under conditions incompatible with life.

That great expert on meteorites Vernadskii wrote as follows:

> *Those germs of life, 'microzoa', cannot have any connection with meteorites or any cosmic dust known to us. For nowhere in the structure*

of the meteorites or dust do we see manifestations or effects of life. If we study them we find that they were formed under conditions similar to those under which our own deepest formations originated (high pressure and high temperature) or else by chemical processes from liquids and gases, also at high temperatures (chondrites, moldavites). Microbes may be associated with them fortuitously but are quite independent and not directly connected with them.

Thus the only possibility would be that the microbes might be picked up by the meteorites in space, but they would then certainly be on the surface of the meteorites and would therefore necessarily be destroyed in transit through the Earth's atmosphere.

A very bold and original hypothesis has fairly recently been put forward by L. Berg. It is directly connected with the meteoritic theory of the transport of life. Berg bases his hypothesis on O. Shmidt's meteoritic theory of the formation of the Earth. According to this theory, the Earth was never an incandescent sphere but consisted of cold materials from the beginning. "Along with the aggregation of meteorites of which it is formed", Berg wrote, "the Earth may also have acquired the germs of life or perhaps ready-made complex living organisms."

This hypothesis, however, agrees so badly with the facts so far studied that it is hard to point to a single fact which might support it. On the contrary, all that we know about meteorites and cosmic dust is totally opposed to it.

Summing up all that has been said, we must admit that the theory of cosmozoë or lithopanspermia, the theory that life arrived on Earth inside meteorites, is in direct contradiction to the objective facts of contemporary science.

Arrhenius' Theory of Panspermia

The theory of radiopanspermia was produced at the beginning of the twentieth century to replace that of lithopanspermia. The originator of this theory was the famous Swedish physical chemist S. Arrhenius, who was an ardent supporter of the idea that life is distributed throughout space. He tried to prove by direct calculations that it is possible for particles of matter to pass from one heavenly body to another. He considered that the main agent in this case would be the pressure of the rays of light.

The phenomenon received its theoretical foundation at the hands of Clerk Maxwell in the second half of the nineteenth century, but the scientists of that time refused to accept it without direct experimental evidence. Only a brilliant experimentalist like the Russian physicist P. Lebedev could succeed in demonstrating the phenomenon, which he did in 1900. By direct experiment Lebedev showed that light exerts pressure on those objects on which it falls and, furthermore, he determined the magnitude of this pressure.

It turned out to be infinitesimal. The sunlight falling on the surface of the Earth only exerts a pressure equivalent to o.5 mg/m2, but even this is enough to cause minute particles of dust to move through a vacuum at a considerable speed.

The experiment of Nichols and Hull which demonstrates the theory well, they used a glass vessel shaped like an hour glass. In it they placed a mixture of emery and very fine carbon dust obtained by the carbonisation of fungal spores. The air was evacuated from the vessel. The stream of particles falling through the narrow opening was illuminated by a powerful source of light. The emery fell to the bottom but the carbon particles were diverted on to the walls.

Arrhenius drew a picture of the passage of small particles, among them the spores of micro-organisms, through interplanetary and interstellar space. Upward currents of air, which would be specially strong after volcanic eruptions, might carry particles of matter to very great heights, up to 100 or more kilometres above the surface of the Earth. In the upper layers of the atmosphere there are, for a number of reasons, constant electrical discharges which would be more than enough to drive these particles of matter out of the atmosphere of the Earth into interplanetary space. Here the particles would travel further and further under the one-sided pressure of the rays of the Sun.

As from the surface of the Earth so, in the same way, very small particles must be constantly becoming detached from the surfaces of other heavenly bodies. If a planet is inhabited by living organisms, particularly micro-organisms, then their spores would be able to travel through interstellar space in the same way. Arrhenius calculated that bacterial spores having a diameter of 0.0002-0.00015 mm could travel through space at a very great speed under the influence of the pressure of sunlight. Fourteen months after having left the Earth such a spore would pass out of our planetary system, but it would be 9,000 years before it reached the nearest star, Centauri. The migration of spores can, however, take place towards the Sun as well as away from it. While wandering in interstellar space the germ may meet comparatively large particles of cosmic dust. If the spore becomes attached to a particle having a diameter of 0.0015 mm it will begin to move towards the Sun, as the pressure of the light will not be able to overcome the weight of the particle which will be approaching the Sun under the influence of gravity. Arrhenius thought that the Earth might have been colonised in this way by spores of micro-organisms coming into our solar system from other parts of the universe.

According to the calculations of Arrhenius the particles of cosmic dust falling on the Earth in this way would not necessarily get hot and burn in the atmosphere of the Earth as do meteorites. If the particles were of the size mentioned, the pressure of light would check their motion and the speed at which they fell would be slow enough for them only to be heated through some tens of degrees, which would not prevent the spores from retaining their viability.

Arrhenius' theory received wide attention in the scientific world and found many supporters both among physicists and among biologists. In the U.S.S.R. in particular it was supported by S. Kostychev, P. Lazarev, A. Nemilov and others. In fact, Arrhenius made careful enough calculations and a good analysis of the mechanical aspect of the passage of particles of matter from one heavenly body to another. There remained, however, the unsolved problem of whether the germs of bacteria could accomplish such an interstellar journey and remain alive. To this aspect of the matter Arrhenius and the other supporters of his theory quite naturally paid special attention.

The distance separating one planetary system from another is tremendous. Even if the particles were to travel at the speed already mentioned it would still be many thousands of years before they reached the nearest star. Under these circumstances one must take into consideration all the dangers to which the germs of life would be submitted during the whole course of their long journey, the severe cold of interstellar space, the complete absence of moisture, oxygen, etc. Could they endure all these hardships for thousands of years while still retaining the ability to multiply when they fell on a new planet, and to give rise to all the later inhabitants of that planet?

The state of the problem at the present day.

The adherents of panspermia expended much work and ingenuity to prove the possibility of such a passage of the germs of life from one heavenly body to another in a viable condition. The spores of bacteria are, in fact, extremely stable under all sorts of unfavourable external conditions. Many of them certainly do not need oxygen. It is well known that anaerobic bacteria can not only be conserved without oxygen but can live without it for the whole of their lives.

In the absence of water due to partial, or even more so to complete, drying, living processes are brought to a standstill but the organism is not by any means always killed. It only goes into a state of anabiosis. This is generally known in the case of the seeds of plants and even such lower animals as rotifers, tardigrada and eelworms. The extensive literature concerning this question is collected in P. Shmidt book *Anabiosis*. The spores of bacteria are particularly resistant to drying. At the beginning of the century L. Maquenne showed that it is even possible to keep absolutely dry seeds in a vacuum for many years and that under this treatment they do not lose their viability. This was later confirmed by P. Becquerel and a number of other authors.

The resistance of bacteria and their spores to low temperatures appears to be exceptional. R. Pictet pointed out this peculiarity of bacteria in the nineteenth century. P. Becquerel kept ampoules containing the dried spores of moulds and bacteria in a vacuum at the temperature of liquid air for several weeks. They all remained alive and grew for a year and a half under observation. The articles of C. B. Lipman and E. Kadisch may also be referred to.

The studies of B. J. Luyet and his colleagues are of particular interest. These studies show that if protoplasm is frozen deeply and quickly with liquid air or hydrogen it is possible to avoid crystallisation of ice and the dispersal of molecules and disturbance of structure associated with it. The protoplasm gets into a glassy state (becomes vitrified) and can be kept in that form at low temperatures indefinitely without losing the ability to be brought to life again when transferred to favourable conditions. From this one may conclude that the germs of bacteria which exist in interstellar space, where the temperature is near to absolute zero, could certainly float around for thousands of years without losing their viability. We find in the literature some reports of the survival of viable bacteria for very long periods in the frozen state, but not all of these reports seem completely reliable. We must refer first to the work of V. Omelyanskil. He found many kinds of micro-organisms (which grew on broth media) in the tissues and mucus of the preserved middle part of the trunk of the Sanga Yurakh mammoth, which was sent to him from the place where the animal was found. The author does not exclude the possibility that some of the bacteria found in the corpse of the mammoth had reached it later. He considers that the evidence in favour of the microflora of the trunk being of contemporary origin with the mammoth is more convincing. If this is true, these bacteria have retained their viability during continuous refrigeration for tens of thousands of years. It must, however, be borne in mind that the remains of the mammoth were sent to Omelyanskii from a distance and were not removed by professional microbiologists. One cannot, therefore, exclude the possibility that they were secondarily infected.

The same applies to the observations of P. Kapterev. He has drawn up a complete list of algae, fungi, bacteria and even crustaceans which he has succeeded in bringing to life from samples of frozen subsoil obtained from a depth of two to seven metres. This implies growth after 1,000 to 3,000 years of refrigeration. L. Kriss studied the frozen subsoil of Kolyuchin and Wrangel Islands and made some very cautious inferences. Although he too found viable micrococci at these levels he considered it perfectly possible that these had fallen there from the upper levels where they were also present.

Thus the problem of the possibility of micro-organisms being preserved in a viable state at low temperatures for thousands of years cannot be considered to be conclusively solved. Nevertheless, one cannot reach the opposite conclusion that bacteria and their spores would necessarily be destroyed at temperatures near to absolute zero.

It seems, however, that the greatest menace to bacteria and their spores in outer space is not so much the cold as the radiations which pass through it. Even at the end of last century it was established that by no means all the radiations of which sunlight is composed reach the surface of the Earth. Part of the light is absorbed by the atmosphere. This absorption affects particularly the ultraviolet radiations which are invisible to the eye but are very active chemically. Only radiations having a wavelength of not less than 3,000 Å reach the surface of the Earth. It is only by going up high mountains that one can establish the presence of ultraviolet light with a wavelength of 2,900 Å. All the short-wave radiation is absorbed by the atmosphere and does not reach the surface of the Earth. However, outside the atmosphere, interplanetary and interstellar space are penetrated by radiations having wavelengths of 1,000-2,000 Å.

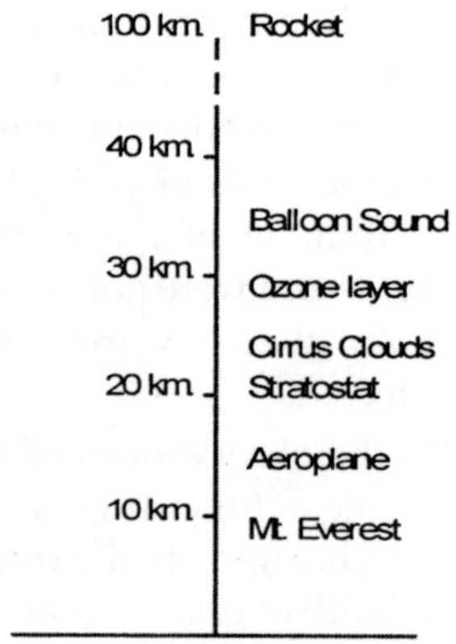

Fig. : The levels of the atmosphere.

These radiations are chemically extremely active. On reaching the outer layers of the atmosphere they are absorbed by molecular oxygen, as a result of which the oxygen is converted into ozone. At a height of about 3 kilometres above the surface of the Earth there is a layer of ozone in the atmosphere called the 'ozone screen' which shields us from the short-wave radiations of interplanetary space. It was noticed as long ago as 1877 that sunshine has a harmful effect on many bacteria. It was later established that this effect is mainly due to the ultraviolet part of the spectrum which has a wavelength of less than 3,100 Å. Using artificial ultraviolet light from a mercury lamp, it was shown that the bactericidal activity of ultraviolet radiations increases as the wavelength decreases. It reaches a maximum at a wavelength of about 2,700-2,800 Å, and then falls off somewhat till the wavelength is about 2,600-2,400 Å, after which it again increases strongly on passing to still shorter wavelengths. In the course of a few minutes, or even seconds, light of this sort will destroy not only the bacteria known to us, but also their spores.

Arrhenius knew about the bactericidal effect of sunlight but he considered that it was not the light itself that killed the bacteria but the oxygen which had been activated by it. This idea seemed to be fully confirmed by the experiments of Roux and Duclos, who kept spores in glass test tubes without oxygen under intense illumination for months. A considerable proportion of the spores retained their viability under this treatment.

These experiments suffered from a technical fault in that all the ultraviolet radiation was absorbed by the glass walls of the test tubes. The experiments of P. Becquerel were technically sounder. He dried the spores of moulds, bacteria and other micro-organisms and collected them on a glass slide which was placed in a wide test tube. This was then hermetically sealed at the top by a plate of quartz, and then evacuated and plunged into a vessel containing liquid air. The spores were then irradiated with a mercury lamp through the quartz plate. They were all destroyed after fairly short periods of exposure.

The supporters of panspermia brought forward numerous objections to these experiments. It was suggested that there are forms of bacteria which are specially resistant to ultraviolet light; that the bactericidal effect of the ultraviolet light is due to oxidative or other chemical changes so that it can only manifest itself in the presence of water and oxygen (these are absent in outer space); that the intensity of the radiations was less in space than in the experiments; that ultraviolet radiation was not effective at temperatures near to absolute zero, etc.

These objections, however, did not stand up to strict experimental testing. At the present time direct experiments using rockets which are sent up to heights of 100 kilometres, i.e. considerably above the ozone screen, have shown that, at this level, ultraviolet radiation is far more intense. We can deduce a curve relating intensity with wavelength for the ultraviolet radiation at the limit of the atmosphere of the Earth.

In his review D. E. Lea also presented a wide range of material showing that all forms of microbes and spores which have been investigated in this respect are destroyed by the action of short-wave ultraviolet light. We now possess considerably greater factual material but it completely confirms the earlier work on the destruction by ultraviolet light of all forms of micro-organisms whatever their species.

Thus the earlier findings of R. Wiesner that there exist forms of bacteria which are resistant to ultraviolet light were not confirmed by later workers. On the contrary, it is now asserted that the various species only differ very slightly from one another in their resistance. This effect of light is quite different from that of temperature, for we know many very thermostable bacteria. This difference is particularly noticeable where spores are concerned. Thus, for example, the spores of *Bacillus anthracis* and *B. subtilis* are very resistant to high temperatures and, in contrast to their vegetative forms, can even undergo more or less prolonged boiling. However, the difference in resistance between the vegetative forms and spores does not exist in respect of the effect of ultraviolet light, which destroys both forms of these organisms almost equally easily.

Another difference between the effects of temperature and light is that the presence of water is not necessary for the effect of light. It has now been established that completely dried cultures and spores of various microbes always exhibit considerable radiosensitivity. Neither does oxygen seem necessary for the bactericidal activity of light. The earlier view that the effect of ultraviolet light depended on an oxidative activity seems to be untenable. It has been shown experimentally that short-wave radiations can have a destructive effect on micro-organisms even in the absence of gaseous oxygen in the surrounding medium. Ultraviolet radiation is bactericidal by virtue of its direct action on the substance of the bacteria.

Neither does the temperature play a decisive part in the process with which we are concerned. F. Gates showed that the temperature coefficient does not exceed 1.06 in such processes, which is as expected for photochemical reactions.

As we have seen, the experiments of P. Becquerel demonstrated the bactericidal activity of ultraviolet radiation even at the temperature of liquid air. This has been confirmed many times since then. In this connection the recent experiments of E. Graevskii are of special interest. This author was studying different forms of bacteria, moulds, yeasts and other such organisms. He showed that when they have been cooled to very low temperatures and the protoplasm is in a glassy state it retains its viability for a long time because, under these conditions, there is no need for metabolic processes to maintain its dynamic structure. However, even under these conditions, micro-organisms and their spores are quickly destroyed by ultraviolet and â-radiation. Graevskii writes:

> *The effect of ultraviolet radiation on a living substrate is the same at room temperature and at 192° C and this completely justifies one in assuming that even the very low temperature prevailing in outer space could not protect living protoplasm from the harmful effects of radiant energy.*

The bactericidal effect of short-wave ultraviolet radiation is explained by its extremely strong chemical effects. The energy of this radiation is so great that it can alter or even disrupt any organic molecules which absorb it. It polymerises acetylene, anthracene and many other hydrocarbons. It decomposes acetone and various aldehydes, organic acids, etc. The effects of such radiations on proteins are particularly interesting to us.

A. D. McLaren has summarised the work of a number of authors in his review. Proteins are denatured under the influence of ultraviolet light and when this happens they lose their solubility in water, they change their viscosity, their optical rotation and their content of amino and other functional groups. In contrast to the denaturation caused by heat, this alteration may occur even on irradiation of the protein in the dry state. Its occurrence is independent of the presence of oxygen. These changes in the physical properties of protein solutions which occur during irradiation (changes in viscosity, solubility, etc.) depend on chemical and structural alterations in the actual molecules of the protein occurring under the influence of the light. These changes are particularly marked at wavelengths where the absorption by proteins is particularly intense. It is specially significant that the curve for the absorption of ultraviolet radiation by proteins corresponds closely with the curve for the destruction of bacteria by radiation in different regions of the ultraviolet spectrum. Thus, in both cases there are maxima at about 2,700 Å; below this, the absorption by proteins and the bactericidal activity fall off and then again increase when the wavelength of the radiations becomes still shorter. This correspondence serves as a clear demonstration that the changes in the protein which are brought about by the ultraviolet radiation are the same as those which destroy the bacteria. It seems significant that direct investigation of irradiated micro-organisms shows that their proteins have been coagulated.

From what has been said it is clear that all micro-organisms which have proteins as the main constituent of their protoplasm (and we know of no living thing which is devoid of protein) must be destroyed by the action of ultraviolet light. As the alteration in the proteins and the associated destruction of the bacteria proceed even in the absence of water and oxygen and at very low temperatures, the probability that viable germs arrived on the Earth from space would seem to be zero. The light of the stars is rich in ultraviolet radiation. On the surface of the Earth we are protected from its harmful effects by the atmosphere surrounding us. On escaping from this atmosphere the germs of life would inevitably be destroyed by the activity of the ultraviolet radiations which traverse interstellar space.

It is true that other 'hypotheses' have been brought forward of recent years in an attempt to redeem the theory under discussion. For example, it has been suggested that life might have been brought here at some time by the landing of astronauts, that is to say, highly developed conscious beings who could undertake interplanetary journeys. This sort of suggestion is, however, more reminiscent of science fiction than of a serious scientific hypothesis. The facts which are at present available to science convince us of the absolute impossibility of viable germs travelling to the Earth through space.

It is interesting to note that, in spite of his ardent belief in the possibility of interplanetary travel, the outstanding Russian scientist and inventor K. Tsiolkovskii nevertheless categorically denied the possibility of this sort of artificial transport of microbes. When he died in 1919 he left a manuscript entitled *The origin of plants on the terrestrial globe and their development*. In it we may read "My work has shown that it will be possible to devise means whereby any living thing may be artificially transmitted from the Earth to another planet and back safely, but mankind is not proceeding very fast towards the realisation of this possibility." However, he goes on to say that this form of transport of life 'with the help of reason' could not have occurred, for no traces had been observed suggesting that at any time or place there have been such highly developed beings deliberately visiting the Earth. Tsiolkovskii wrote in conclusion: "This means that life did not reach the Earth from the planets even with the help of reason."

5

Body Structure and Posture

Introduction

A vast majority of the visible and gross features of the vertebrate body are related to the function of locomotion. Primitive locomotion in fish was achieved by a wavelike movement of the vertebral column in a lateral plane. The trout moves its tail from side to side, and the waves of muscle contraction pass forward up the animal's body. In primitive fish the fins are used mainly as stabilizers and for steering, but in later forms they may be used for propulsion, especially the paired fins: the *pectoral* fins behind the head, the *pelvic* fins nearer the tail.

The first fish to climb from the water used their pectoral and pelvic fins as props for their bodies, and thus came to initiate the process whereby the locomotor function was eventually transferred from the vertebral column to what was to become of the limbs. The paired *pectoral* and *pelvic* fins were attached to the backbone or vertebral column by three paired bones of the same name, later to become the pectoral and pelvic girdles. The lateral muscles at the sides of the vertebral column (the *myotomes*, which affected the wave motion of the body) were then greatly reduced, and the limb musculature developed in their place. An intermediate stage can be seen in the newt, which uses both its myotome muscles and its limbs for propulsion.

The vertebral column assumed a new function, that of supporting the body above the limbs. The flexibility so necessary for swimming was replaced by a stiffness necessary to carry the head and body. The soft parts already *ventral* to (on the underside of) the vertebral column in fish became suspended from it in vertebrates, and the head became supported by a cantilever. The resulting structure was compared by D'Arcy Thompson to that of a cantilevered bridge and can be understood as a weight-bearing compound girder of webbed construction. The important muscles are now ventral and *dorsal* to (that is, below and above) the vertebral column no longer lateral. This important change in the evolution of vertebrates is clearly demonstrated in the secondarily aquatic whales, which evolved from terrestrial quadrupeds and which propel themselves in the sea by moving their tails up and down rather than laterally.

In the terrestrial quadruped, therefore, the vertebral column is under compression the *body* of each vertebra (and the *intervertebral discs*) sustains the load, and the *supraspinous* ligaments are under tension, while the *nuchal muscles* of the neck support the head. The bending moment (that is, the force tending to cause bending in the vertebral column because of the weight of the body and head) is reflected in the length of the *spinous processes*, which, together with the oblique *interspinou sligaments* and muscles, constitute the web of the compound girder. The spinous processes are rarely so greatly developed over the pelvic girdle as over the pectoral because in most vertebrates the weight of the tail is less than that of the head, though exceptions to this are known. Further stressing of the center section of the girder between the limbs is achieved by an inverted parabolic girder between the limbs of which the *sternum* and abdominal muscles form the ventral tension member.

Bridges, however, are designed for rigidity, and the anatomy of bridges is only comparable to that of animals insofar as animals require rigidity in their skeletons. Flexible structures are of a different order of complexity and do not so clearly follow the simple mechanical principles of the bridge-builder. That is why the cantilever structure is seen more clearly in the large slow-moving animals than in the small, active forms. Flexibility means that the bones must be jointed to allow movement, and the spinous processes of the vertebrae must be short enough not to touch each other when the body is flexed. As a result, the girder is supported by ligaments under very great stress; it has gained flexibility at the cost of strength.

The second factor to be recalled when studying skeletal architecture is that when the loads due to gravity are very slight, as in small animals, the structures associated with locomotion will mask those designed to withstand the force of gravity. In large animals, like the bison or the dinosaur, the structural postural elements will be more strikingly developed than the locomotor elements.

We shall trace the evolution of posture from the simple quadrupedal plan with horizontal trunk to the species with erect trunks that, as we have seen, are found among prosimians as well as many other higher primates, including humans. The vertebral column, which began in fish as a flexible rod under intermittent lateral compression (when the myotomes contract) is, in quadrupeds, the compression member in a weight-bearing compound girder. In primates it eventually becomes a vertical weight-bearing flexible rod, held erect by muscular "rigging," like the mast of a yacht. We shall see that this change of function is accompanied by an appropriate change in form.

Center of Gravity

Differences in the form of the vertebral column among quadrupedal animals are due to differences in posture, in locomotion, and so in the center of weight of the animal, better known as the center of gravity. This center will move forward or backward along the backbone depending on whether there is more body weight in the head and shoulder region or in the hindquarters and tail region. In the bison, the majority of weight, as in most terrestrial quadrupeds, lies over the forelimbs, and that is where the center of gravity lies. They have been described as forelimb dominant. In contrast, the dinosaur and kangaroo, for example, both have far more weight over the hindlimbs, and that is the feature which allows them semierect posture with tail support, so that they can gather food with their forelimbs. The hindlimbs are dominant.

A somewhat analogous adaptation can be seen in all the primates regardless of their limb proportions. Here the center of gravity is well back for a number of reasons. The first is that the musculature and length of the hindlimbs has increased, especially in the leaping and grasping prosimians, to give extra locomotive power: the hindlimbs are dominant. The second is that the tail has enlarged in many groups into an organ for maintaining balance when jumping. It is also necessary that, when an animal is propelled forward, the force of propulsion should pass through the center of gravity; otherwise the animal would revolve in the air. It therefore follows that a center of gravity near the organ of propulsion, in this case the back legs, will add to the stability of the jumping primate. It is noteworthy that in this feature all the primates contrast with almost all other quadrupedal terrestrial mammals (except the leaping cats), which have forelimb dominance.

The energy required for locomotion is also a factor related to the position of the center of gravity. It appears that at small body sizes a posterior center of gravity is energetically advantageous, while at larger body sizes an anterior position is more efficient. Most primates are relatively small animals, so this factor again supports the evolution of hindlimb dominance. The fact that the only fully terrestrial large primate (*Homo sapiens*) is bipedal is also significant, because this adaptation has been shown to be more efficient than that of a (theoretical) large hindlimb-dominant quadruped.

It is for these reasons that in the early stages of primate evolution we see the center of gravity well back over the hindlimbs. This development makes practical the sitting position and frees the forelimbs for manipulation, just as we see in the squirrel and kangaroo. It also makes possible the key adaptations of human evolution, when the earliest hominids became able to stand and run on their hindlimbs. The forest adaptation of a low center of gravity over the hindlimbs was a necessary prerequisite for this final evolutionary development. In human evolution, while the tail was lost, the center of gravity moved further down as a result of development of the hindlimbs, somewhat reduced musculature in the forelimbs, and some changes in the form and position of the trunk and abdomen.

In living humans the whole skeleton is modified so that the body's weight can be carried with minimum effort and maximum stability. This means that most of the weight is transmitted by bones with the shortest possible cantilevers. The line of weight passes from the occipital condyles down just in front of the vertebral column to the upper sacral vertebrae, then through the pelvic bones and down the hindlimbs to the tripods of the feet. As we shall see, the bioengineering of the human body is unique in the animal kingdom.

Evolution of the Vertebral Column and Thorax

Adapted as many of them are for arboreal locomotion, the higher primates have evolved very flexible vertebral columns compared with most quadrupedal plains mammals. Flexion of the backbone has been added to that of the legs to give extra propulsive power in jumping, as well as to absorb the shock of impact. Flexibility was achieved by some small modifications of the vertebral column, which are most clearly seen in the quadrupedal forms:

1. There are well-developed intervertebral disks that act as elastic buffers and allow tipping movement of the vertebrae.
2. The spinous processes are relatively short, with the result that extensibility is not restricted by their proximity.
3. The *thorax* (the rib cage and its musculature), which is deep and which gives the kind of support to the trunk that is achieved by a parabolic girder, stiffens the vertebral column along the thoracic region. In contrast the *cervical* (neck) and *lumbar* (waist) regions of the vertebral column in quadrupedal monkeys are relatively long, and these regions in particular allow bending movements of the body.

Table 1. Average Numbers of Thoracic and Lumbar Vertebrae.

	Thoracic vertebrae		Lumbar vertebrae	
Primate	**Mean**	**Range**	**Mean**	**Range**
Papio	10	9-11	9	9
Hylobates	13	12-14	6	5-7
Great apes	13		4	
Humans	12	10-12	6	5-7

In climbing and brachiating forms, however, this flexibility is somewhat lost, since the vertebral column plays a far smaller part in locomotion. We find, for example, that three of the nine lumbar vertebrae found in quadrupedal monkeys are absent in gibbons, and that the lumbar region is reduced in all the great apes. It is, however, interesting that the stiffening has proceeded further among the great apes than it has in humans. As a whole, these data suggest strongly that we evolved from a primate that engaged in arboreal climbing rather than one with a history of quadrupedalism.

But other fundamental changes have clearly occurred in the spine and thorax during human evolution that were adaptations both to a history of erect posture and climbing and to the later evolution of bipedal locomotion:

1. The size of the vertebral body increases at the pelvic end of the vertebral column because the forces of compression are no longer constant along the trunk as they are in a quadruped with a horizontal backbone. As the force of compression increases from head to pelvis because of the weight of the animal, so the cross-section of the vertebral body, which bears it, must also increase. This increase is present in the great apes but is greatest among humans. The proportions of the vertebral body also change, and the weight distribution on the vertebrae themselves changes in order to carry the vertical forces of compression.
2. The spinous processes are more or less equally developed along the vertebral column because the bending moment the tendency to bend is no longer restricted to the points of support, the limb girdles, but is now more evenly distributed. For example, the gorilla has well-developed cervical spines that anchor the nuchal muscles, but they are reduced in modern humans. As a result of our fully erect stance and bipedal locomotion, the head, although heavy, is more or less balanced on top of the vertebral column.
3. Among all the climbers, the inverted parabolic girder effect of the sternum and abdominal muscles, which stiffens the backbone, becomes less important. The sternum tends to lie nearer to the vertebral column, its segments become fused and broadened, and the form of the thorax is wider and shallower, bringing the center of gravity nearer the vertebral column. Schultz has studied the angle between the "neck" and "body" of the ribs and found that among the monkeys the angle varied from 89 to 100 degrees, while among the more erect apes and humans it varied from 50 to 65 degrees.

   A wider but flatter thorax also gives greater power to the lateral abdominal muscles. Ventral displacement of the vertebral column may similarly improve the leverage of some of the back muscles (e.g., iliocostalis). Finally, it has the effect of allowing increased shoulder mobility, since it alters the position of the shoulder girdle, as well as moving the shoulders apart. It is clear that a wide shallow thorax is advantageous to an arboreal climber.

4. Increased weight transmission through the pelvis and legs has resulted in an enlargement of the *sacrum*, the zone of fused vertebrae that:

   We see from these data that we have a relatively stiff backbone, which suggests that our origin lies among climbers, and we no doubt benefitted from this preadaptation in the evolution of a fully erect bipedal posture. In achieving the posture of a climbing species and later a bipedal one, the vertebral column evolved to form a vertical weight-bearing towerlike structure in place of a compound girder, which means (as we have seen) that the vertebral bodies, which transmit the weight to the pelvic girdle, show an increase in size as we pass down the vertebral column, and the spinous processes, which supported the webbing of the girder, have become much reduced. The thorax has flattened to allow the center of gravity to move back toward the vertebral column, which now effectively lies within the thoracic area. The weight of the body is transmitted to the legs via the pelvic girdle.

The evidence of human ancestry derived from the vertebral column is important and is supported by the evidence of the fossilized vertebrae of *Australopithecus*, which show most of the humanlike features associated with an upright posture. However, the vertebral column and rib-cage still carried some ape-like features, the bodies of the lumbar vertebrae are still relatively small, and the spines on the cervical vertebrae appear to be longer than in *Homosapiens*. These observations confirm what we know about the general course of human evolution.

Head and Neck

Our purpose here is to consider the head in its relation to the rest of the body and the cervical vertebrae in particular. Just as semi-erect and erect postures have affected the form of the *cervical vertebrae*, so have they also affected the position of the head in relation to the spine and the connecting musculature.

We have seen, in such quadrupedal forms as the bison or monkey, how the head is supported by a webbed girder consisting of the cervical vertebrae under compression and the nuchal muscles and ligaments under tension. It is clear that as a more erect posture evolved, less strength was required in the nuchal muscles as well as in the spinous processes of the cervical vertebrae, since more of the weight was carried directly by the bodies of the vertebrae.

At the same time, we have already mentioned that in primate evolution generally the muzzle was reduced and the brain enlarged. This development resulted in the center of gravity of the head itself moving more nearly over the point of pivot of the head upon the vertebral column. We clearly see the center of gravity moving back until, in humans, where the masticatory apparatus is also reduced, it is almost exactly over the occipital condyles. It has not yet achieved a perfect balance and probably will not do so, because there are no powerful ventral muscles present to support the skull from the front and stop it from tipping backward. Those that are ventral to the occipital condyles the prevertebral muscles -arise very close to the occipital condyles and so exert only slight leverage on the head.

Le Gros Clark has proposed a useful means of assessing the extent to which this movement of the center of gravity has occurred in different primates: by measuring the position of the occipital condyles in relation to the rest of the skull. The *condylar position index* is calculated according to the simple geometrical basis and some figures for this index are recorded. Note that the increase of the index has just begun in *Australopithecus*, but has not at this stage progressed very far in the human direction. The index numbers may be compared with the drawings. This double change in posture and in the head's center of gravity has had the effect of removing work from the nuchal muscles, which have in turn been reduced. Schultz has shown that taking size into account, six times as much power is needed in the nuchal muscles of monkeys and apes than is needed to support the human head. This reduction in musculature has been reflected in the area of insertion of the nuchal muscles upon the back of the skull. When we compare, for example, the gorilla and modern humans, the difference is striking. The fossil skull of *Homoerectūs* is intermediate in this respect. Like the Neanderthal people that followed somewhat later, *Homo erectus* also has a well developed masticatory apparatus, a long head, clearly defined nuchal area, and well-developed nuchal musculature. The balance of the head was less perfect that it is in modern humans.

The balance of the head, and the power required in the nuchal musculature, is also reflected in the development of the spines of the cervical vertebrae. We have atlas and axis fragments, and a sixth or seventh cervical vertebra from Hadar; the latter carries a long spinous process. The *Homo erectus* boy from Nariokotome has an almost complete vertebral column which is very similar to that of modern humans. However, the surface areas of the vertebral bodies were small for his weight, the spinous process of the cervical vertebrae long, and the vertebral canal which carries the spinal cord was relatively small. It has been suggested that this implies a less precise control of movement of the thoracic cage which generates breath control for language. We have no other fossil hominid cervical vertebrae until we come to the Neanderthal people; here again, the spines are significantly longer that those of modern *Homo sapiens*. The final reduction of the spinous processes appears to have been a very late development.

Table 2. The Position of the Occipital Condyles.

Primate	Mean condylar position index
Proconsul africanus	30
Pan gorilla	24
Pan troglodytes	23
Pan paniscus	35
Australopithecus africanus	39
Homo sapiens (modern)	77-81

The reduction of the nuchal musculature has affected the form of the skull, for the area of attachment in humans in not only much reduced, but it is no longer bordered by a clearly defined nuchal crest. The back of the skull is relatively smooth and follows more closely the outline of the brain, for in humans the outer layer of bone of the skull (the outer "table") responds very little to the much weaker nuchal muscles during growth.

A characteristic of the base of the skull that is related to its balance is the orientation of the foramen magnum the opening through which the spinal cord passes from the brain. In quadrupeds, the foramen faces more or less backward, but in modern humans it has come to face more or less vertically downward. This feature was noted by Raymond Dart in 1925 when he described the first *Australopithecus* skull. It suggested to him that the skull had belonged to a bipedal animal.

It has been mentioned that there are no powerful ventral muscles in the neck to keep the head from tipping back. This presents a problem probably unique to humans, because the center of gravity of the head is well forward in all other primates. We can tip our head back beyond the point of balance, and we must be able to pull it forward again. The forward movement of the head is made possible not only by the weak prevertebral muscles mentioned above, but also by the *sternomastoid muscles,* which, lying on each side of the neck and inserted upon the *mastoid process* of the skull and the sternum and clavicle at the top of the thorax, are individually able to rotate the head upon the vertebral column. Their importance to the primates is great, since this group is dependent on rotary head movement to maintain effective allaround vision. When in the very last phase of human evolution the center of gravity of the head moved back near the occipital condyles (the pivot), the sternomastoid muscles could just counteract the backward forces upon the head by contracting together.

Krantz has shown that the sternomastoid muscles obtain effective leverage to pull the head forward only because an enlargement and lengthening of the mastoid processes has occurred in human evolution. This lengthening alone makes it possible for the sternomastoids to raise the head after it has tipped back; only by lengthening is the insertion point of the muscles brought forward beyond the point of pivot when the head is thrown back. As it is, the leverage is very poor, a fact that can be verified by throwing your head well back over the back of your chair and slowly raising it again, with your hands on your sternomastoids. The enlargement of the mastoids in human evolution is thus a direct result of our erect posture and of the forward movement of the occipital condyles in relation to the skull as a whole.

In this section we have seen how we come to have a reduced nuchal musculature, a reduced nuchal area, and elongated mastoid processes. All these developments are facets of a double adaptive complex that (a) involved the adoption of a fully erect posture, and

(b) involved changes in the olfactory region, the masticatory apparatus, and the brain, which together moved the head's center of gravity and the relative position of the occipital condyles. All these changes moved the center of gravity of the head to a position only slightly ventral to the position of the occipital condyles themselves.

The Tail

As we have seen, the tail is used by many primates for balance in jumping, and has evolved in such a way as to be one factor in bringing the center of gravity back toward the hindlimbs. It is also used in many groups as a rudder and elevator (such as the tail of a bird or airplane), to adjust the position of the body as it passes through the air so that the animal will arrive with its limbs in a position to grasp on impact. This function accounts for the thick hair on the tail of most jumping primates, which increases its aerodynamic effect. We have also noted that in the New World monkeys the tail has evolved as a grasping organ a fifth limb adding to mobility as well as to stability.

The tail is also used in other ways by terrestrial monkeys, among which it tends to be reduced. It acts as a support for the baby riding on its mother's back like a jockey, as a device for signaling (as in dogs), as a flywhisk for the anogenital region (as in cattle), or it may be almost completely lost, as in the stump-tailed macaque. Among macaques alone it varies from 27 to 3 vertebrae.

As we have seen, the apes and humans the Hominoidea have lost their tails. Reduction has gone furthest in the gibbons: as compared with 20 to 30 caudal vertebrae found among most monkeys, the gibbon has only one or two. The apes as a group are tailless, and this seems to be a product of their climbing mode of locomotion, even if the gibbons also brachiate. In a climbing species the tail serves no aerodynamic or other purpose and has been lost. The only apelike creature with any evidence of a tail is the very early primitive genus *Aegyptopithecus* from the Oligocene of Egypt, which was much smaller and probably much more active than modern apes. It is also of interest that the slow-climbing (SQC) prosimians, such as the lorises, are also tailless.

In humans the tail is also reduced, but we retain from three to five small caudal vertebrae known as the *coccyx*. At least six are present in the human fetus, but three of these fuse together during growth. As this reduction is shared with all the Hominoidea, we can attribute it to an ancestral condition which we believe was that of an arboreal climber. The adoption of bipedal locomotion did not require any major reversal of the condition, though some modification seems to have occurred.

It is of interest that reduction of the tail in modern humans has not progressed as far as in the gibbon. This may be because the coccyx has assumed a totally new function arising from our fully erect posture. The tail vertebrae, the coccyx, are now curved ventrally and help form the basinshaped structure that carries the viscera. Ligaments run from the forwardpointing coccyx to the *ischium* the lower part of the pelvis.

At the base, the sack is rounded off by the curved bones of the pelvis, the sacrum, and the coccyx, as well as the *sacrotuberous* and *sacrospinous* ligaments. This viscera-carrying function of the pelvis is again something new and is associated with bipedal locomotion.

6

Structure and Biological Functions of Proteins and Nucleic Acids

Introduction

In the beginning the word albumen was only applied to the substance in hens' eggs which forms a white coagulum when heated. Later on, other substances similar to the white of eggs were included in the term albumen, but this concept was not given any general biological significance in relation to life. On the contrary, it was considered that egg albumen and other analogous substances were no more than the specific products of a few isolated organisms and, in particular, that they were completely absent from plants. Thus, for example, the gluten which had been isolated from flour as early as the end of the eighteenth century was regarded as a curiosity, a freak of nature, and even called *matière végéto-animale*. However, as the study of the chemical substances of living nature proceeded, so the idea became stronger and stronger in the minds of scientists that albumens are present in all organisms and that these compounds play an extremely important part in the process of life. This idea received precise expression in the name given to albumens in the 1830s by G. J. Mulder. He called them *protein*, (first or most important). In using this term Mulder was thus stressing the biological aspect of protein as the most important component of living material. At that time chemical knowledge of proteins was very meagre. Proteins attracted the attention mainly of biologists, who usually regarded them as the main and most important components of the gelatinous material within the cell. This material was called 'protoplasm' by H. v. Mohl in the middle of the nineteenth century and the part it plays as the material carrier of life became more and more evident. Some biologists of the latter half of the nineteenth century even identified protoplasm with protein and among them E. Haeckel, for example, considered that the simplest organisms consisted of nothing but lumps of proteinaceous substances. F. Engels, in common with the biologists of his time, often used the terms 'protoplasm' and 'albuminous bodies' (*Eiweisskörper*). The 'proteins' of Engels must therefore not be identified with the chemically distinct substances which we have now gradually succeeded in isolating from living things, nor with purified protein preparations composed of mixtures of pure proteins. Nevertheless Engels was considerably in advance of the ideas of his time when, in speaking of proteins, he specially stressed the chemical aspect of the matter and emphasised the significance of proteins in metabolism, that form of the motion of matter which is characteristic of life.

It is only now that we have begun to be able to appreciate the value of the remarkable scientific perspicacity of Engels. The advances in protein chemistry now going on have enabled us to characterise proteins as individual chemical compounds, as polymers of amino acids having extremely specific structures. As well as this we can to a certain, though admittedly very limited, extent relate this structure to enzymic and other biologically important properties of proteins. This will enable us to understand their extremely great significance in the metabolic process of life. Many organic substances of different kinds entering into the composition of living protoplasm can only readily take part in its metabolism after they have interacted with the proteins of the protoplasm to form extremely active complexes (enzyme-substrate complexes). In the absence of such interaction the chemical reactions of which these substances are capable take place too slowly at ordinary temperatures for them to have any significance in the rapidly moving process of life. Hence the metabolic course followed by any organic compound will depend not only on the peculiarities of its molecular structure, its chemical potentialities, but also on the specific enzymic activity of those proteins of the protoplasm with which the compound is involved in the general metabolism.

Thus, in proteins (enzymes) living material has both powerful catalysts to accelerate chemical processes and an internal chemical apparatus whereby these processes are directed along completely determinate paths co-ordinated with each other in a definite sequence and forming the orderly arrangement of processes characteristic of metabolism. On the basis of this organisation there also takes place, in particular, the constant regeneration of proteins, their self-reproduction, by virtue of which, to use Engels' words, the protein body "while being the result of ordinary chemical processes, is distinguished from all others by being a selfacting, permanent chemical process. "

This presentation is, of course, radically different in principle from those hypotheses formulated at the end of the nineteenth century which identified protoplasm with proteins and referred to the so-called 'living protein molecule'. In these hypotheses, which were discussed more fully in Chapter 3 of this book, some workers attempted to treat protoplasm as a whole, as a single chemical substance, as a gigantic protein molecule endowed with life. Others regarded protoplasm as no more than a specific medium, a mixture of lifeless compounds containing the hypothetical living particles, the protein molecules, in the chemical structure of which there lie concealed all the causes and mysteries of life. We may refer here to the 'biogens' of M. Verworn, the 'moleculobionts' of Alexander and Bridges and other similar hypothetical particles, the chemical reality of which has never been proved by anyone, though references to them are still to be met within scientific literature.

Thus contemporary chemists and biologists use the word protein' in a long series of different senses. At one end of the series we have the purely chemical definition of proteins as highly polymerised organic compounds with very complicated molecules made up of different sorts of amino acids. This definition would, however, seem to be very onesided. It ignores the biologically important properties possessed by all the various proteins which have actually been isolated from organisms, properties which are related to the individual peculiarities of their structure. Such a definition would include all polymers of amino acids, even such possible combinations of amino acids as would not subserve the biological functions proper to naturally occurring proteins. Polymers of amino acids of this sort would naturally be unable to form part of the structure of living matter. This purely chemical definition, therefore, includes among proteins even substances which have no direct biological significance. On the other hand, the definition which we find at the other end of our series, that of the living protein molecule, is completely lacking in any clear-cut chemical meaning. The partisans of this concept attribute to the protein molecule (in most cases they refer to molecules of nucleoproteins) all the properties of life, i.e. the ability to metabolise, reproduce themselves, etc. However, they give absolutely no real explanation of how all these properties could depend on any particular arrangement of the atoms in the hypothetical 'living molecule'.

As a result of this confusion, many contemporary authors studying the origin of life make quite arbitrary and illogical jumps between the concepts of protein implied by the purely chemical and the purely biological definitions. For example, they argue as follows: if the process of organic-chemical transformation in the waters of the primaeval ocean could have given rise to protein-like polymers of amino acids, then the same processes must have led to the formation of 'living protein molecules'. In what the specific 'life-conferring' structure of these molecules consists and how it could have arisen seems to be something of an inessential detail from this point of view; this structure might even have been formed as a result of purely fortuitous combinations of groups of atoms which remained unchanged during the reproduction and multiplication of these molecules in all succeeding generations. The perpetrators of arguments of this sort do not, however, notice that their approach to a solution of the problem in hand is purely formal and verbal in character and that what they regard as a detail constitutes the very essence of the question.

It seems to us that the problem of the primary development of proteins should be formulated in a different way, as follows: the numerous and varied proteins which we can now isolate from living organisms in crystalline form as individual chemical compounds (various enzymes, hormones, viruses, etc.) have definite structures which are highly specific to each of them and which are extremely well adapted to the fulfilment of those vitally important functions which they subserve in living protoplasm (in metabolism, in reproduction, etc.). Substances of this kind only arise nowadays as components of living bodies and there can be no doubt that the specific structures which they now exhibit reflect the earlier evolution of these bodies and are the result of the prolonged development of living organisms.

The main point of the question is whether compounds of this kind could arise outside living material, primarily. To give a satisfactory answer to this question it is necessary to give at least a short account of what is now known of the chemical structure of the actual proteins which have been isolated from living things and to try to understand which are the specific features of their structure responsible for their biologically important functions. Only after this shall we be in a position to reconstruct for ourselves the ways by which there arose, during the process of the development of matter, those structural peculiarities of the primaeval polymers of amino acids which are required for the vital processes. In discussing the chemical structure of proteins we must first make clear to what extent these 'working mechanisms' of protoplasm which have been isolated from living organisms (various enzymes, hormones, toxins, etc.) exist at the molecular level and to what extent they appear as chemically definable substances, in connection with which the concept of a molecule is the same as for other organic compounds. As early as 1940 N. W. Pirie expressed doubts as to the validity of this approach and to some extent these doubts still appear in the scientific literature on proteins.

In fact, many proteins which were earlier thought to be individual substances have been shown, by more refined methods of separation, to be mixtures. For example, egg albumin has been shown to be a mixture, notwithstanding the fact that it forms beautiful crystals. The same is true of serum globulins. For many years purified casein was considered as a single protein. This seemed to be proved by the good agreement of the analytical results obtained by scientists in different countries. However, it has now been established that pure casein consists of a mixture of at least three proteins which have been separated from one another.

In his detailed paper dealing with the isolation of proteins J. F. Taylor points out what a complicated matter it is to obtain individual proteins from naturally occurring mixtures of them. At the end of his paper he gives a list of those proteins which are now recognised as chemically homogeneous compounds. We cannot be certain, however, that even these proteins are completely uniform.

In connection with the lack of molecular homogeneity of casein, G. R. Tristram has also pointed out that â-lactoglobulin is not a single substance either, and rightly poses the question as to whether the proteins which are now held to be individual substances are not really mixtures of related compounds, among which even the amino acid composition varies somewhat. Certainly there are a number of facts which suggest that several pure individual protein-like substances may form, as it were, a family of proteins, being composed of the same amino acids but differing from one another in the amounts of some of the amino acid residues in the peptide chain. This may be demonstrated particularly clearly as regards haemoglobin.

In this connection we must emphasise the fact that proteins having the same biological function may differ markedly from one another chemically. Insulin serves as a good example of this. The hormone was isolated from the pancreas as an individual protein of comparatively low molecular weight, the structure of which is now very well worked out. However, it has been shown that the insulins obtained from oxen, pigs and sheep, though they have the same physiological activity, nevertheless differ from one another chemically. In particular, pig insulin contains threonine at a position in its peptide chain where it is not present in ox insulin. Thus it is evident that the physiological properties of hormonal proteins do not require absolute uniformity of structure. The same may also be said of enzymes. It now seems quite clear that we include under the same name (pepsin, invertase, phosphomonoesterase, etc.) proteins which have the same enzymic activity though they sometimes differ markedly among themselves in respect of molecular size, isoelectric point and other physico-chemical properties and even in respect of their amino acid compositions.

It follows that the catalytic properties of a given protein are not associated with the whole of its molecule and that this may contain parts which are completely inactive and can easily be altered without destroying the enzymic properties. It follows that some variations in amino acid composition do not necessarily cause noticeable alterations in their biological properties.

It is now well known that different forms of organisms can contain proteins which are identical in their biological functions but which differ in their amino acid composition. It has also been established that changes in the living conditions of organisms bring about variations in the composition and properties of their proteins.

Having made a thorough review of the facts which we have referred to, Tristram draws from them the following conclusion: "That proteins do appear to remain more or less constant in composition may well be a reflection of the constancy of an environment, rather than evidence that proteins are compounds of unvarying composition."

The amino acid composition and sequence in the structure of the macromolecules of proteins.

Having made these indispensable remarks about proteins as individual chemical substances we can now proceed to a proper description of the fundamentals of protein chemistry. It may now be held to be firmly established, in the first place, that protein molecules are made up of residues of various amino acids and, in the second place, that these residues are linked together in the protein molecule mainly by peptide bonds between the α-amino groups and α-carboxyl groups of amino acids, as was first suggested by A. Ya. Danilevskii and afterwards proved experimentally by E. Fischers and F. Hofmeister and a number of later workers.

Thus, as a first rough approximation, a protein molecule may be described schematically as a polypeptide chain:

This sort of structure fundamentally distinguishes proteins from other organic polymers such as cellulose or rubber, in the molecules of which the same atomic grouping (residues of glucose, isoprene, etc.) is repeated over and over again.

Thanks to the variety of amino acid residues entering into their composition, and also to the great chemical variety of their functional groups, proteins have enormous chemical potentialities. They can react with the countless multitude of substances of living protoplasm to form either true compounds of the nature of conjugated proteins or extremely ephemeral complexes which only have a very transient existence, as happens in the formation of intermediate compounds (enzyme-substrate).

Arising from this, many students of proteins from H. Ritthausen to present-day authors (e.g. H. B. Vickery and W. H. Stein) have put forward the suggestion that the chemical, and even the physiological, characteristics of any particular protein could be deduced from a detailed and complete knowledge of its amino acid composition and an understanding of the properties of the different amino acids of which it is made up.

Quantitative and qualitative analytical studies on various proteins with a view to determining their amino acid composition have been going on for many years. However, the methods devised in the classical works of A. Kossel, E. Fischer and T. B. Osborne and others depended on the separation of amino acids from hydrolysates and involved the expenditure of enormous amounts of effort, time and starting materials. For this reason such studies were very few and far from complete. However, there have been introduced into protein chemistry in recent years new and satisfactory micromethods based on up-to-date principles of investigation, (isotope dilution and the isotope-derivative method, microbiological assay and chromatography). This led to signal advances in the field of amino acid analysis and a very large number of proteins may now be taken to have been fully analysed in this respect. (The extensive factual material is given in the numerous tables in the article by G. R. Tristram.)

Detailed studies have also been made of the chemical properties of the separate amino acids which are found in proteins, those which are common to all carboxylic acids and primary amines and also the specific functional attributes which belong to each separate amino acid and characterise its radical (R). The extensive data on this subject have been recently collated in a review by P. Desnuelle.

The results obtained in this way were, however, rather unexpected. In particular, it was found that only a very limited number of different amino acids are to be found in natural proteins, especially in those of higher animals and plants. It is interesting to note in this connection that, as the number of substances studied grows greater and the accuracy of the results improves, the number of so-called 'common' amino acids found in proteins does not increase; in recent years it has even shown a certain tendency to decrease. Thus H. B. Vickery and C. L. A. Schmidt considered, in 1931, that there were twenty-two common amino acids, this number later fell to twenty-one, and now, as P. Desnuelle writes, "we shall therefore assume twenty common amino acids only, and this number will probably never be much modified".

The question inevitably arises as to why the endless variety of proteins which we can isolate from contemporary animals and plants should be made up of such a limited number of structural elements. It is clear that as a result of the physical and chemical laws discussed in the previous chapter there could and must have been formed many, many other amino acids as well as those given in the list. Why, then, do we not find them in contemporary proteins? Obviously, in the formation of these latter, there must have occurred a strict selection of those amino acids indispensable for life. It would seem that the chemical functions which we have just discussed are quite sufficient for the catalysis and co-ordination of all the various metabolic reactions. Combined with one another into protein molecules the twenty amino acids listed form all the enzymes necessary for metabolism and the other important internal chemical mechanisms of living protoplasm. There can be no doubt that in the process of evolution of living matter there took place a rationalisation of these mechanisms and consequently a standardisation of them, analogous to that occurring in technical processes. All those amino acids which were not absolutely necessary to life were eliminated in later generations by natural selection.

The idea of the standardisation of the amino acid composition of proteins during the process of evolution of higher organisms finds support in the fact that among organisms at a lower stage in evolutionary development mainly bacteria and fungi we find, in addition to the ordinary amino acids which are constantly present in proteins, that there are continually being discovered new, so-called 'peculiar' or 'uncommon' amino acids such as α-thiolvaline in *Penicillium* spp., meso-α-diaminopimelic acid and other amino acids. There are also found in the proteins and peptides, and particularly in the antibiotics, of lower organisms the 'unstandardised' D-forms of amino acids, among them Dglutamic acid in the capsule of *Bacillus anthracis* and related organisms, D-leucine in gramicidin, D-phenylalanine in gramicidin S and tyrocidine (*B. brevis*), D-alanine in *Lactobacillus arabinosus*, etc. In higher organisms, on the other hand, we invariably find only L-forms of amino acids and apparent exceptions to this rule have always been found to be artefacts arising by racemisation, usually during the hydrolysis of the proteins.

Thus we see that during the course of evolution the transition from the lower forms, with their as yet imperfectly organised metabolism, to higher forms in which the metabolism has reached a higher degree of co-ordination, is marked by a standardisation of the amino acid composition of proteins due to natural selection. Thus, one of the essential properties of the animal and vegetable proteins which we have studied, their amino acid composition, is not entirely determined by physical and chemical laws alone but carries the imprint of its biological origin.

Another deduction which can be drawn from a thorough study of the amino acid composition of present-day proteins does not bear out the optimistic expectations, referred to above, of many chemists of the past and present centuries beginning with Ritthausen and ending with Vickery. It has been shown that even the most complete amino acid analysis of a particular protein taken by itself is still far from characterising the physical and chemical properties of that protein, let alone its biological functions. As K. Bailey wrote recently:

> *One of the most disheartening features of the amino-acid analysis of proteins is that the results have little meaning. To a limited extent they are useful for assessing the nutritional value of a protein, but they do not explain at all the true biological function; why one protein is an enzyme, another a hormone, another a toxin.*

This is quite understandable even on purely theoretical grounds. Never in the history of science could it be maintained that the whole is nothing but the sum of its component parts. ("The whole is always somewhat different from the sum of the separate parts," M. Planck, 1935.) In proteins it is the structure which determines this difference. Even a study of the chemical properties of artificially synthesised polypeptides shows that the chemical activities of free amino acid groups (the so-called radicals, R) are markedly changed when they are included in peptide chains, and also depend on the order in which they are arranged in the chains.

Even by simply comparing the effects of various substances and physical factors on a mixture of amino acids and a polypeptide composed of those same amino acids, it will be found that the amino acid residues of peptides are considerably more labile. For example, they are far easier to racemise than the corresponding free amino acids. In just the same way, the chemical reactivity of particular functional groups such as the hydroxyl group of serine, the phenolic hydroxyl group of tyrosine, the ù-amino group of lysine, etc., is very substantially altered according to which chemical groups are immediately adjacent to them in the polypeptide chain. In a number of cases amino acid radicals, when forming part of a polypeptide chain, can react with compounds to which they would be quite indifferent in the form of free amino acids. G. R. Tristram sums the matter up as follows:

> *It is now appreciated that the properties of side chains in a protein are not simple functions of the properties of the free amino acids, but are, in fact, highly complex functions dependent on many factors including the relative distribution of side chains in the main peptide chain and in the folded native protein.*

The order in which the amino acid residues are arranged in the peptide chains of native proteins or biologically important peptides isolated from organisms has long attracted the attention of scientists. The elucidation of this problem has, however, been attended by a very large number of technical difficulties.

The first reasonably successful attempt to establish the order in which all the amino acid residues are arranged in a single protein-like substance was made by K. Felix and his colleagues on the protamine of herring sperm, clupeine. They considered that the molecule of clupeine is made up of no more than 33 amino acid residues, namely, 22 of arginine, 2 of alanine, 2 of serine, 3 of proline, 3 of valine and 1 of hydroxyproline. The unusually small number of amino acid residues and absence of any great diversity, in particular the extreme predominance of arginine, greatly simplified the task of studying this peculiar protein. In Felix' opinion the amino acids in the polypeptide chain of clupeine are arranged in a rather regular way; there are always a series of four arginine radicals in a row along the chain followed by two residues of other amino acids and then again four arginine residues, etc.

M. Bergmann put forward the hypothesis that the polypeptide chains of other proteins besides clupeine are constructed similarly and that they also contain a definite repetitive sequence of amino acid residues. For example, if there are 54 lysine residues in edestin out of a total of 432, this means that every eighth amino acid residue in the polypeptide chain of edestin will be lysine. Similarly, in ox globin, each of the 36 lysine residues will be separated from the next by 15 residues of other amino acids and each of the 12 residues of proline will be separated from the next by 47 other residues, etc.

Hormones, Enzymes, Antibiotics and Antigens

The structure is fundamentally related to the ratios between the numbers of different amino acid residues in any particular protein was not confirmed by the direct study of the breakdown products obtained by partial hydrolysis of proteins and polypeptides. On the contrary, the application of this tedious but very reliable method has actually enabled people to elucidate the very complicated arrangement of the amino acid residues in the peptide chains of a number of physiologically important compounds. This applied, in the first place, to toxic substances produced by bacteria (antibiotics), such as gramicidin. By partial hydrolysis of the simplest of these, gramicidin S, which is a cyclopolypeptide, it has been possible to isolate numerous dipeptides and tripeptides by paper chromatography. By comparing these, the whole sequence of amino acids in this cyclic peptide has been established. Furthermore, the use of similar methods has led to the elucidation of the sequence of the amino acid residues in tyrocidines A and B and other polypeptide antibiotics.

Corresponding studies on proteins are naturally of special interest. The one which has now been most thoroughly studied is insulin. This hormone, which is particularly important on account of its physiological activity, has been studied by many workers, both in respect of its amino acid composition and in respect of the arrangement of the amino acid residues in the polypeptide chain.

The work of F. Sanger and his colleagues has given a clear picture of this structure. Sanger marked the terminal amino groups of the polypeptides present in insulin by condensing them with 2:4-dinitro-1-fluorobenzene; he then submitted this derivative of insulin to partial hydrolysis and studied the breakdown products obtained in this way. On the basis of the results thus obtained Sanger then arrived at the structure of insulin as follows: the molecular weight of soluble insulin is about 48,000. It varies, however, with the concentration and pH of the solution. When the pH is less than 4 or more than 7 the insulin molecule dissociates into parts with a molecular weight of 12,000. These parts are each composed of four open polypeptide chains in two of which (the A chains) the terminal amino group belongs to a glycine residue and the terminal carboxyl group to an aspartic acid residue. The corresponding terminal residues in the other two chains (the B chains) are phenylalanine with a free amino group and alanine with a free carboxyl group.

Sanger and colleagues consider that, strictly speaking, the basic unit of insulin is a particle with a molecular weight of 6,000 consisting of one A chain and one B chain joined together by disulphide bridges. According to C. Tanford and J. Epstein," two such particles are joined together by means of zinc atoms to form a particle with a molecular weight of 12,000.

These data as to the sequences of amino acid residues in the polypeptide chains of insulin do not show the periodicity in the arrangement of amino acids suggested by Bergmann. The arrangement here is far more complicated. Two identical amino acid radicals may be side by side or may be separated from one another by any number of other residues. There is no obvious regularity or rhythm in these sequences. Moreover, a definite sequence must be present, at least in some part of the molecule, if the protein is to exercise its physiological functions. We still do not know why this is so, we cannot explain the immediate cause of this specificity, but facts which have been obtained recently demonstrate beyond doubt that the specificity exists both for insulin and for other analogous hormones. For example, the two hormones of the posterior lobe of the pituitary (oxytocin and vasopressin) are very similar in their amino acid composition, but even the small differences in the details of their structure confer on each its own essential hormonal function which it alone can carry out.

In oxytocin the sequence is Cys. Tyr. Ileu. Glu. Asp. Cys. Pro. Leu. Gly. In vasopressin the sequence is Cys. Tyr. Phe. Glu. Asp. Cys. Pro. Arg. Gly.

Interesting results have been obtained by P. H. Bell and R. G. Shepherd concerning the structure of the ß-adrenocorticotrophic hormone (ACTH). This is a polypeptide with a molecular weight of 5360 and the following arrangement of amino acid residues: Ser. Tyr. Ser. Met. Glu. His. Phe. Arg. Try. Gly. Lys. Pro. Val. Gly. Lys. Lys. Arg. Arg. Pro. Val. Lys. Val. Tyr. Pro. Asp. Gly. Ala. Glu. Asp. Glu. Leu. Ala. Glu. Ala. Phe. Pro. Leu. Glu. Phe.

It is important that the biological activity of the hormone only depends on the presence in the correct order of the amino acids 1 - 24 starting from serine. The rest of the chain is of no importance for its hormonal activity.

Unfortunately we have not yet got the same information concerning the sequences of amino acids in enzymes that we have in hormones. In this case, however, the connection between these sequences and the specific catalytic action of the enzyme in question seems clearer. If the enzyme is to hasten the transformation of the substance which acts as its substrate it must first combine with that substance.

For enzymes with two components which have specific nonprotein (prosthetic) groups as well as proteins in their molecules, it has long been established that the combination of the enzyme with the substrate takes place through the prosthetic groups.

For example, W. Langenbeck showed in his model experiments that the ability of carboxylase to catalyse the decarboxylation of pyruvic acid is associated with the presence of an amino group in the molecule of the enzyme. Simpler compounds containing this group, such as methylamine, can also accelerate this reaction. The catalytic activity of methylamine is, however, very slight, but can be made many times greater by incorporating, in the molecule of methylamine, carboxyl, phenyl and other groups which do not in themselves have carboxylase activity but considerably augment this activity of the amino group.

It has further been shown that natural carboxylases, that of yeast, for example, have as a prosthetic group a phosphorylated derivative of vitamin B_1. In this compound the amino group is combined with the heterocyclic pyrimidine and thiazole rings, which confer on it a very high catalytic activity. Not only that, but when the thiamine pyrophosphate is combined with a specific protein the complex acquires a catalytic activity as a decarboxylase nearly 10,000 times greater than that of the most efficient of Langenbeck's artificial models.

7

Manipulation and the Forelimb in Life Science

Introduction

In the locomotion of the terrestrial quadrupedal mammals, including the quadrupedal primates, the main driving force is derived from the hindlimbs. The body is propelled forward from the rear (not pulled), and mechanical efficiency requires rigid joints between the pelvic girdle and the vertebral column. The function of the forelegs, even in a generalized mammal, is different. They carry the majority of the weight, and in movements such as running and jumping, usually hit the ground first and take a great deal of shock, which would be transmitted throughout the body by a rigid bone-to-bone connection. Such a connection does not exist; the shock is absorbed by the mass of muscle that suspends the spine and thorax from the shoulder girdle.

Although the shoulder or *pectoral* girdle consisted in its primitive condition of three bones, as does the pelvis, it is reduced in mammals to two the *scapula* and *clavicle*. The scapula is the dorsal bone and is equivalent to the ilium; the clavicle is ventral and equivalent to the pubis. Unlike the pelvis, however, they are not fused but are connected by a moveable joint (*acromioclavicular joint*). The clavicle makes contact with the rest of the skeleton through a second movable joint, the *sternoclavicular*, where it articulates with the *manubrium* or upper part of the *sternum*. The clavicle maintains the distance of the scapula from the sternum and, acting as a strut, allows movement of the scapula at a constant radius from the manubrium. The weight of a quadrupedal animal is suspended on the *serratus anterior* muscle. The clavicle keeps the *glenoid cavity* (the pectoral equivalent of the acetabulum), into which the *humerus* fits, at a fixed distance from the midline of the body.The clavicle is not an essential component because movement occurs only in the anterior-posterior plane, and there is no lateral swing of the forelimbs. The clavicle has been lost in these groups.

The muscle suspension of the forelimbs gives them greater flexibility than the hindlimbs, and this flexibility has been exploited, as we shall see, by the primates.

The intermembral index of the higher primates indicates that the apes have long forelimbs in relation to their hindlimbs. However, an allometric analysis in which forelimb length is related to body size shows that only the gibbon and two New World genera (*Ateles andBrachyteles*) have forelimbs of exceptional length. All these species depend to a large extent on suspensory locomotion, which is often rapid, and have specialized in the evolution of long forelimbs. It is notable that the pattern of elongation is equally true of both components of the forelimb, the humerus and radius.

The Pectoral Girdle

They can move some 40 degrees around the sternoclavicular joint in each of two planes. At the same time, the scapula can move freely in relation to the clavicle around the acromioclavicular joint. These two movable joints allow the scapula to move over the surface of the thorax in both the vertical and the horizontal planes. The acromioclavicular joint allows the scapula to twist in relation to the clavicle and so remain flat on the surface of the thorax, in spite of the fact that the latter is not spherical. The scapula moves forward around the thorax in pushing or thrusting and to the back in pulling the body forward or upward in climbing. In addition, by the action of the *trapezius* and serratus anterior muscles, the scapula can be rotated so that the glenoid cavity (equivalent to the acetabulum in the pelvis) can face in different directions. This rotation nearly doubles the range of movement of the limb in a vertical plane. It may also be noted that the glenoid cavity is more open than the acetabulum and therefore allows a greater arc of movement in the forelimb than in the hindlimb. Here, stability has been sacrificed for flexibility.

From this short account of the pectoral girdle, it can be seen that, in accordance with its different function, its structure is very different from that of the pelvis. In fact, the structure is more generalized that is, capable of more kinds of movement and we might accordingly predict a greater variety of adaptations in the course of primate evolution. Such is indeed the case. More striking differences in the position of the shoulder girdle upon the thorax, differences related to the change of function of the forelimbs in the evolution of the primates.

We can describe three distinct functions of the forelimbs and follow their changing proportions. These functions are: (1) to support the weight of the body in quadrupedal locomotion; (2) to suspend the weight of the body in arboreal climbing; and (3) to manipulate objects.

The change from quadrupedalism to climbing is reflected in the form of the clavicle and of the scapula and its musculature. We find that among primates with well-developed forelimb mobility the clavicle is longer; among quadrupedal species it is shorter. Its greater length in the former alters the position of the glenoid fossa to facilitate suspensory arm movements.

The scapula and its musculature have a more interesting story to tell. The shoulder girdle can be rotated to allow the arm to rise vertically, a movement that is clearly important in locomotion by suspension. Oxnard has shown how the importance of the two rotatory muscles, the serratus anterior and the trapezius, varies among the main locomotor groups of primates. The data indicate that, as might be expected, the muscles are better developed in apes than in quadrupeds and that in this characteristic humans are closer to the apes. At the same time, there is a relative increase in length of the bony lever to which these muscles are attached (as we saw in our consideration of leverage about the hip joint). That is, the insertion area of the muscles on the scapula tends to move away from the central point around which the scapula rotates. Therefore, in species using forelimb suspension we find that the scapula is lengthened; it is longer in the monkeys, especially the arboreal species, and longest in the Hominoidea. The *acromion process*, which, with the ventral bar of the scapula, carries the trapezius and *deltoid* muscles, is also well developed in these groups for the same reason.

The actual plane of the glenoid cavity has similarly changed in relation to the rest of the scapula, reflecting the characteristic position of the arms in relation to the body and the greater mobility of the shoulder joint in species that climb and swing by their arms. Quadrupeds, therefore, tend to have the glenoid cavity pointing more or less horizontally; climbers.

In this angle humans are aligned with the quadrupeds, since the plane of the human glenoid cavity is more nearly parallel to the lateral margin of the scapula than it is in climbing species due to the fact that humans do not regularly raise their arms upward but, on the contrary, carry them hanging down from the shoulder. The human bipedal stance has clearly changed the shape of the scapula in this respect; we need not conclude that the larger angle indicates a history of quadrupedal locomotion.

On the other hand, when we turn to fossil hominid clavicles and scapulae we find a different situation. At present, fossil remains of pectoral girdles are rare. The lightness of the scapula and its position within the surrounding musculature probably render it liable to destruction by predators and scavengers. Clavicles are more common. There are two fragments belonging to the Miocene *Proconsul*, but, apart from their obviously hominoid character, they are too incomplete for further diagnosis. Of *Australopithecus* we have fragments of six clavicles, three from Hadar and one from Makapansgat, all fragmentary, and two from Olduvai. Of the latter, which are quite well preserved, one is described as only slightly different from that of modern humans but suggesting a history of suspensory locomotion. One clavicle is known from Zhoukoudian (*Homo erectus*), and it appears to be quite similar to a modern human clavicle.

There is a fragmentary scapula of *Proconsul africanus* that is described as a diminutive version of a chimpanzee scapula. There is one fragmentary scapula from Hadar and one from Sterkfontein attributed to *Australopithecus*. (There is also a very small fragment from Koobi Fora, which may also belong here.) Fortunately, both consist of the anterior part, including the glenoid cavity and part of the ventral bar, and the Sterkfontein specimen includes part of the *coracoid process.* The coracoid is curved strongly and the bicipital tuberosity at the upper end of the glenoid fossa is well developed: these characteristics indicate suspensory use of the arms and especially climbing. At the same time, the small angle that the plane of the glenoid cavity makes with the ventral bar of the scapula (the *axilloglenoid angle*) again puts both bones into the suspensory class. Broom summarized his description of the Sterkfontein scapula by stating that the bone falls between that of the orangutan and modern humans in its overall form; or, stated another way, it had a more suspensory function than the modern human scapula.

Although they are only small fragments, these scapulae from the opposite ends of Africa may provide an important clue to the locomotor history of *Australopithecus.* Two fragmentary bones constitute a small sample, but they do suggest either that *Australopithecus* used its arms in a suspensory mode for climbing, at least to some extent, or that its recent ancestors did so. Which of these hypotheses is more likely to be correct will be clarified further by the evidence of the skull and our discussion of the ecology and behavior of the group.

We can distinguish, therefore, three types of scapulae associated with three locomotor patterns: the quadrupedal scapula, shorter and squarer in proportion, with a ventrally pointing glenoid cavity; the suspensory or climbing scapula, elongated, with a cranially pointing glenoid cavity; and the human scapula, still elongated, but with a ventrally pointing glenoid cavity. The anatomy of the *Australopithecus* scapula suggests some degree of climbing in the ancestry of hominids, and in the final evolution of *Homo* we see a change in the angle of the glenoid cavity associated with bipedalism.

Evolution of the Upper Arm

The humerus is a powerful bone that carries a considerable proportion of the body weight in quadrupedal mammals; the forward muscles of propulsion, the *pectoralis* and *latissimus dorsi* muscles, are attached to it. Among suspensory species, these muscles are used to lift and support the body; the latissimus dorsi muscles, which are dorsal to the thorax, also help to lift the body in climbing by retracting the arm. Oxnard has shown that they are relatively larger in suspensory and climbing species than in quadrupeds, for only in the former do they lift the total body weight. The arm itself, on the other hand, is lifted by the deltoid muscle, which again is larger in suspensory and climbing species. At the same time, the point of insertion of the muscle on the humerus (the deltoid tuberosity) moves down the bone to give greater leverage.

Less striking differences include the plane of the articular head of the humerus in relation to the elbow joint. This relationship has changed in evolution according to the position of the scapula on the thorax. Thus, in quadrupedal forms the center line of the ball joint of the humerus, which fits into the glenoid cavity, is just about at right angles to the elbow facing directly backward, while in the chimpanzee the humerus has twisted so that the angle between elbow and head is nearer 30 degrees, and the head points inward. The arboreal langur monkey *Presbytis* is intermediate in this characteristic. Humans share this feature with the apes; in both climbing species and bipeds the arm is able to move laterally as well as backward and forward.

Finally, it is characteristic of climbing species as well as humans that the hinge joint at the elbow, the *trochlea*, allows full extension of the forearm and gives considerable lateral stability at that point. In the characteristics of the trochlea, the great apes and humans are very close.

In the upper arm, therefore, we find that humans share two characteristics with climbing species: (1) relatively large arm-raising muscles and a low point of insertion of the deltoid, and (2) a humerus with head pointing inward rather than straight backward. However, our upright stance has led to a reduction in musculature in the climbing muscles, the pectoralis and latissimus dorsi, compared with all other primates and especially the suspensory species.

Let us now turn to the fossil evidence for the evolution of the hominoid forelimb. We have fragments of two humeri of *Proconsul*, over ten of *Australopithecus*, and one of *Homo erectus*. One *Proconsul* humerus is associated with a radius and ulna, and these important fossil bones, which belong to the small species *Proconsul africanus*, have been described in fifteen different studies, including Napier and Davis, Morbeck and Feldesman. The humerus has been well reconstructed and in form and dimensions appears somewhat similar to that of an arboreal monkey. The point of insertion of the deltoid muscle is well down, and the trochlea closely approaches the condition seen in *Pan*, allowing great extension at the elbow joint. It has been suggested that suspensory locomotion was combined with intermittent quadrupedalism. On the basis of this limited evidence we can conclude that *P. africanus* is a small ancestral ape, possibly specialized for below-branch feeding. The overall impression is of a powerful, climbing animal that was not much constrained by joint and ligament morphology to a limited range of postures.

Small portions of two femora and a humeral shaft have also been reported from the chimpanzee-sized *Proconsul nyanzae*. These fragments do not bear clear suspensory characteristics, but their body size and limb proportions suggest they were below branch feeders.

It is not possible to draw more than very limited conclusions from these few fragments of *Proconsul*, but it is fair to admit evidence of small climbing proto-apes in the Miocene *Proconsul africanus*. While we may never know for certain what place this Miocene species had in the evolution of the Hominoidea, at least the limbs do not preclude its close association with a hypothetical Miocene ancestor for both apes and humans.

The proximal end of the humerus bears a few characteristics that may enable us to understand its function. It is impossible to know the angle of the head unless the bone is more or less complete. The development of the tubercles will indicate the importance of the muscles they carry: the lesser tubercle carries the *subscapularis*, a medial rotator and the greater tubercles the *supraspinatus*, the *infraspinatus*, and the *teres minor*, which raise the arm upward and rotate it laterally. The tubercles are separated by the bicipital groove, which lodges the long tendon of the biceps muscle. The Sterkfontein humerus consists of the head and greater part of the shaft and has been described by Broom. In general, he finds the head of the humerus almost typically human, but the lesser tubercle is very well developed and suggests a powerful *subscapularis* muscle an important muscle in climbing. The Koobi Fora specimens stand out as peculiar and distinct from existing Hominoidea; their status is in doubt, but they probably belong to the heavily built *A. boisei*.

Table 1. Sites Bearing Australopithecus Humerus Fragments.

Aramis	Both extremities
Kanapoi	Distal extremity
Sterkfontein	Proximal extremity
Kromdraai	Distal shaft and trochlea with ulna fragment
Olduvai	Distal shaft and trochlea
Koobi Fora	Parts of eight humeri (2 well preserved)
Hadar	Parts of nine humeri (4 well preserved)

The trochlea and distal end of the humerus is easier to interpret but far less varied among hominoids than the proximal end and cannot be easily used to distinguish human from ape. The earliest hominid find of such a fossil is the fragment from Kanapoi, interpreted as morphologically close to *Homo* by McHenry and Corruccini, but close to the Hadar finds and, surprisingly, the robust fossils from Koobi Fora, as interpreted by Feldesman. It seems to be an early *Australopithecus* of rather robust build. It is notable that, like modern humans, it lacks the trochlear ridge that stabilized the elbow joint of the African apes. The Aramis and the Hadar specimens are slightly closer to modern humans but not much. The distal humerus fragments from Koobi Fora are of later date, are very robust, and are not closely related to the other specimens. They possibly belong to the little known robust *A. boisei*. All the fossil specimens are typically hominid, however, and quite distinct from all other primates including *Hylobates* and the monkeys. The trochlea unites humans and their ancestors closely to the African apes.

The scapulae and arm bones from Nariokotome have not yet been described in detail but appear to be modern in aspect. A humerus shaft from Zhoukoudian associated with the fossils of the Beijing population has been described. Though unusually thick-walled, this bone appears to fall within the human range. The specimens of femora and humeri are few, yet it seems clear that *Homo erectus* had limbs not dissimilar to our own. If it were not for the evidence of the skull, we might be tempted to assess these limb bones as belonging to a robust race of *Homo sapiens*.

From the data presented here it seems clear not only that the human arm carries characteristics associated with climbing primates, but also that our fossil relatives, *Australopithecus*, show strongly the characteristics of an ape that frequently elevated its arms in climbing. With the minor exceptions noted, the total morphological pattern conforms closely to that seen in the African apes. The evidence for climbing locomotion in human ancestry accumulates.

The Forearm

The brachial index was devised as an indicator of the relative lengths of the humerus and radius and is still sometimes used. It is better, however, to use a logarithmic display of the proportions of the forelimb. The proportions are constant, though in humans and in the gorilla some shortening of the forearm relative to the upper arm has occurred. The reasons for this shortening are not understood, though in the case of the gorilla it is possibly associated with that heavy animal's more quadrupedal mode of locomotion: enormous weight is transmitted by the forelimbs. Keith has suggested that shorter human forelimbs would have resulted in more precise movements of the hand around the elbow.

The bones of the forearm, the radius and ulna, are flexed and extended about the hingelike elbow joint by muscles that originate on the humerus and lie along it. The *triceps brachii* is the extensor of the forearm; the *brachialis* and *biceps* the flexors. As might be expected, the extensor is the important muscle in a quadrupedal mammal because it supports the weight of the animal at the elbow, and the flexors are important in climbing forms because they flex the elbow and lift the animal. Oxnard has again shown that the relative mass of the triceps is, as we might expect, diminished in climbing species. The length of the power arm of the forearm lever (that is, the distance of the *bicipital tuberosity* from the elbow joint) also varies with locomotor function. In this characteristic, humans lie between the quadrupedal and the suspensory condition.

The retention of the two bones in the forelimb is due to the need for effective rotation of the hand (compare tibia and fibula), called *pronation* and *supination*. In pronation the thumb lies nearer the body; in supination digit 5 does so. The articulation of the elbow joint is primarily between humerus and ulna, and at the wrist between the radius and the carpal bones. Forces of compression and tension are transmitted between the two forearm bones by the *interosseous membrane*. The extent of rotation is limited by the extent to which the radius and ulna are bowed. While the supination of quadrupedal monkeys is about 90 degrees, that of apes and humans is nearer 180 degrees. The power of rotation is dependent on the mass of the *pronator* and *supinator* muscles. There is, again, a correlation between extensive rotation of the forearm and arboreal climbing, as might be expected.

When the arm is fully extended, the *olecranon process* of the ulna fits snugly into the *olecranon fossa* of the humerus, so that the forearm cannot extend beyond 180 degrees. This is an important characteristic in quadrupedal species that take weight on their arms. However, it is interesting that among the African apes, especially the gorilla, the olecranon process is reduced. This appears to allow overextension of the forearm, which would give greater than usual stability in a heavy animal that has taken up quadrupedal knuckle-walking. In contrast, the human olecranon process is more protuberant than in apes. This has been interpreted by Tuttle and Basmajian to be a result of the use of tools and weapons, which required powerful and rapid extension of the elbow joint.

Fossil bones of the forearm are extremely rare. Some from *Proconsul africanus* suggested a relatively short forearm, but new discoveries suggest otherwise. In most characteristics of the forearm, the *Proconsul* fragments show an intermediate condition between quadruped and climber, as we have seen in earlier sections.

There are a number of fragments of radii and ulnae from Hadar on which no comparative studies have yet been made. A small part of the proximal end of a radius from Koobi Fora has been described as carrying a well-developed bicipital tuberosity. A proximal fragment of a small right ulna is also known from the same skeletal collection. Comparative studies are awaited. A very well-preserved large ulna from Omo has been attributed to *Australopithecus boisei*. It is difficult to interpret but has characteristics which are present in both *Pan* and *Homo*.

We have three fragments of radii of *Australopithecus* from Makapansgat and one from Swartkrans. There is a proximal ulna fragment from Kromdraai. Only the ulna has been studied in detail, and all are claimed to be similar to those of modern humans. Fragments of the forearm bones belonging to Neanderthalers though robust and powerful are within the variation of form found among living *Homo sapiens*.

The Hands of Primates

The *manus* of the primitive tree shrew is a simple, fivefingered (*pentadactyl*) organ, the rather short phalanges making it appear like a paw. It is not prehensile, nor are the phalanges opposable. Each terminal *phalanx* carries a claw, and there are six volar pads of friction skin on the palm and one terminal pad on each digit (finger). Just above the wrist lies a small skin projection (*papilla*) from which long hairs grow, the *carpal vibrissae*. This papilla is a tactile sense organ common to many mammals, and similar to those that bear whiskers on the face in dogs and cats. The vibrissae are implanted in richly innervated skin and form an extremely sensitive but undiscriminating tactile sense organ.

Studies of comparative anatomy make it clear that, as the volar pads increase in area and sensitivity in primate evolution, the importance of the carpal vibrissae decreases. They are poorly developed in the New World and absent in the Old World anthropoidea. In the human fetus, a small transient cutaneous papilla can sometimes be seen in the carpal region. During evolution the innervation of the volar pads and particularly of the terminal digital pads increases in density so that they become tactile sense organs of great importance.

While it happens that in humans the tactile function of the hand completely replaces the locomotor function, the manus remains an important locomotor organ in all other primates. The forelimb may not be as important as the hindlimb in the propulsion of the quadrupedal species, yet as a grasping organ it is always important. This grasping function is enhanced among the climbers.

All primates, including humans, possess prehensile hands and convergent-divergent fingers. The coming together of the fingertips (convergence) and their spread (divergence) result from the arched form of the *carpus* (wrist) and *metacarpus* (palm), for the fingers are set in a curve at the wrist. Prehensility is a very important characteristic of primates, since it permits one-handed feeding. The squirrel, which lacks prehensility in its manus, has to use both forepaws to feed.

There are two means of achieving a hold on a branch. The first, which is typical of the prosimians, involves a *pollex* (thumb), which diverges from the phalanges and can be opposed to them around the branch. The second, which is typical of suspensory species, especially brachiators, involves the lengthening of the phalanges and, in many cases, a reduction of the thumb. In the quadrupedal Old World monkeys, however, the opposable thumb is retained as a result of the evolutionary development of the hand for use in other functions. These include the procurement of food, grooming, hygiene, defense, the care and carrying of offspring, improved three-dimensional tactile sense, and so forth. A baboon has a precision grip between thumb and fingers sufficiently well evolved to extract the sting from a scorpion and feeds by plucking individual grass stems by hand: manual grazing.

In certain arboreal monkeys and some Hominoidea we find the thumb much reduced, since the more the hand is involved in suspension, the less it is available for manipulation. This specialization reaches its most extreme form not in the gibbon, where the thumb is retained for limited manipulation, but in *Ateles* and the African *Colobus* monkey, where it is vestigial or absent.

A second characteristic of climbers and brachiators is the hooklike hand with its long phalanges. The proportions of the hands of apes have been investigated by Schultz. When hand length was compared to body trunk length, it was found that while in monkeys hand length is usually less than one-third of trunk length, it rises to an average of 47% in New World monkeys and as much as 59% in the gibbon.

The evolution of an opposable thumb is a feature of great importance in human evolution. The thumb reaches its greatest relative size among prosimians, such at *Indri* and in some prosimian species the second digit is much reduced to permit an even more effective grip on a branch. Although in these forms the thumb is completely opposable to the digits, Napier have described this condition as *pseudo-opposability* to distinguish it from the somewhat different human condition. In the latter, we share with the Old World monkeys and apes a modified kind of carpo-metacarpal joint, which gives the thumb a different range of movement and results in *true opposability*.

Among the suspensory species of monkey and great ape with true opposability, the thumb is relatively reduced from a longer primitive condition. The great apes *Pan* and *Pongo* have such short thumbs that they cannot achieve the powerful pad-to-pad contact of thumb and other digits that forms the anatomical basis of humankind's precision grip.

Table 2. Mean proportions of the Hand and Thumb in Primates.

Primate	Hand breadth x 100 Hand length	Pollex x 100 Hand length
Presbytis	29	40
Macaca	35	56
Pan troglodytes	34	47
Pongo	34	43
Pan gorilla	51	54
Hylobates	20	52
Homo	43	67

Apart from the prosimians, therefore, a fairly long thumb can be found only among the quadrupedal Old World monkeys, lesser apes and humans, and, as we have seen, it is widely used in many of these groups for the manipulation of objects.

A fairly complete fossil hand belonging to the Miocene *Proconsul africanus* has been described. It appears to be basically similar to the hand of a quadrupedal Old World monkey with certain signs of suspensory adaptations. The hand is unfortunately incomplete, and in particular it is not known how long the thumb was; nevertheless, its assessment as showing similarity to a suspensory or climbing species seems justified. This conclusion accords with that drawn on the basis of the forelimb bones. It seems clear that in *P. africanus* we have an incipient climbing ape, and one that, on the basis of the limb bones, could conceivably be ancestral to the living Hominidae.

Evolution of The Human Hand

We have seen that monkeys use their hands for a wide variety of purposes, and one of the most important, which we must now consider, is their contribution to the satisfaction of the exploratory drive: the hands make possible a detailed examination of parts of the environment that can be manipulated. Monkeys, apes, and humans are almost the only animals that fiddle about with things, that turn them over and examine their form and texture. This manipulation of objects is not necessarily directly related to the procurement of food, but is simply a process of investigation equivalent to the olfactory investigation of the environment so characteristic of a dog. But in manipulation a monkey is investigating only one particular part of the environment and frequently a part that can be separated physically from the rest. This ability of higher primates to extract an object from its setting and examine it visually and three dimensionally from all sides is a development of the utmost importance in human evolution. A carnivore will examine the olfactory nature of objects but cannot simultaneously see them as part of the visual pattern of the environment, because during examination by the nose the objects are more or less out of sight. Admittedly, a ball or bone can be manipulated with the paws and tossed with the mouth, but the range of objects examined is limited, and, compared with the primate hand, the paws and mouth give only a rough indication of shape and texture.

The examination of things as objects we owe to the evolution of the primate hand and the opposable thumb. The recognition of different kinds of objects we owe to primate visual and tactile examination. Only a primate can, as it were, extract an object from the environment, examine it by smell, touch, and sight, and then return it to its place. In this way the higher primates came to see the environment not as a continuum of events in a world of pattern but as an encounter with objects that proved to make up these events and this pattern. We shall return to this development in considering perception and conceptual thought.

This all-important development may have come quite early in primate evolution, with the appearance of the opposable thumb and the use of the hands in feeding. In higher primates, as we have seen, a further advance in opposability has perfected this remarkable organ of manipulation. Napier writes that true opposability involves "a compound movement of abduction, flexion, and medial rotation at the carpo-metacarpal articulation of the pollex." A saddle joint evolves at the base of the thumb that allows a 45-degree rotation as well as movement in two planes.

Though monkeys and apes have such an opposable thumb, it is only in humans that the thumb is long enough and divergent enough to carry a heavy musculature. This length and strength make possible a precision grip strong yet delicate. At the same time, with the hand no longer functioning to grasp branches, the *metacarpals* and phalanges straighten out in the course of human evolution, and the terminal segments broaden.

Fragments of the hands of *Australopithecus* individuals are known from Sterkfontein, Olduvai, and Hadar, and they are reported to be somewhat less than fully human. The Hadar hand bones constitute the largest collection and have been described by Stern and Susman, who stress their morphological similarity to apes, though Tuttle and Johanson see both ape and human characteristics in them. Starting from the wrist, the two *capitate* bones from Hadar retain a generalized hominoid morphology. Both are waisted, as is the specimen from Sterkfontein and in this feature resemble *Pan* and some arboreal monkeys. The *pisiform* from Hadar, which is rod-shaped, is also different from the human bone (pea-shaped) and resembles those of apes and monkeys. The elongate pisiform is associated with the powerful *flexor carpi ulnaris* muscle in apes and gives increased leverage for flexion of the wrist (carpal) bones that must be important in climbing locomotion.

The Hadar trapezium the wrist bone that carries the thumb is an important bone since it carries the carpo-metacarpal articulation of the pollex; it closely resembles the joint in chimpanzees and is significantly more apelike than the specimen from Olduvai, though even that is less than fully human. The bones of the pollex form a thumb that lies in its relative length halfway between that of apes and that of humans. The first metacarpal of the thumb has a joint that matches the trapezium in its apelike structure, and the bone itself is not so strongly developed as it is in modern humans; the proximal phalanx has the same apelike characteristics.

The remainder of the metacarpals resemble recent finds from Sterkfontein of approximately the same age. They are somewhat curved but contrast with *Pan* in lacking a transverse dorsal ridge and expanded articular surface on the metacarpal head features associated with knuckle walking.

The proximal phalanges are slender and curved as markedly as in a chimpanzee, much more curved than anything seen in modern humans. The muscle markings and joints in all phalanges show that the fingers could flex powerfully, and this implies suspensory and grasping function. They are all generally apelike.

There are no features in the *Australopithecus* hand suggesting a knuckle-walking adaptation or even a knuckle-walking ancestry. In contrast, the hand bones in our possession indicate clearly that the hand functioned as a grasping organ. The thumb shows little evidence of the opposability and strength characteristic of the modern human thumb.

Almost nothing is known of the hand of *Homoerectus* (four damaged phalanges), but it is of interest that although the Neanderthal people had a brain as large as our own, their hands were slightly different and possibly a little less dextrous.

When apes and monkeys carry objects about, their locomotor efficiency is impaired. Tool carrying, food carrying, and food sharing (which is possible only if the food is first carried) all have been reported to a limited extent among chimpanzees, but the animals cannot move quickly with a handful of objects and do so only for very limited distances. In early humans the survival value of being able to carry objects without a disadvantageous effect on locomotion is of the greatest importance. Not only is it most advantageous to be able to carry food when surprised by a competitor or a predator, but this ability can lead eventually to food gathering and so to food sharing. The survival value of carrying also serves in turn to promote a physical structure adapted to bipedal locomotion.

At Koobi Fora, stone tools have been found dated at over 1.8 mya; they consist of battered cobbles, choppers, and flakes, and show an assured technique in their production. At another contemporary site nearby, stone tools are found in association with the bones of a hippopotamus. From Olduvai Gorge, Bed I, we have a succession of increasingly varied tool kits just under 1.8 million years old, often in association with *Homohabilis*, which have been described in great detail. Here we see pebbles, probably first used for pounding roots and smashing bones, preserved after flaking as chopping tools for the preparation of meat. It was probably an essential element in the evolution of a carnivorous diet among hominids.

The tools found associated with and presumably manufactured by *Homohabilis* are relatively simple and crudely made. Finely flaked stone tools are found (to date) only in association with the bones of hominids that had larger brains. Fine Acheulian hand-axes are known from Ternifine as well as Olduvai, where they are associated with remains of *Homo erectus*. Clearly, the final perfection of the hand consisted not only in its anatomy but in sensory perception and motor control associated with a highly developed brain. The most clear-cut neural correlate of the evolution of the human hand is the increase in representation that this organ carries on the motor and somatic sensory cortices of the brain. We have also noted significant improvement in the direct nervous pathways to the hand of the ascending and descending fibers.

We can summarize the evolution of the human hand by marking three probable stages in its development. First, the primate arboreal environment selected the pseudo-opposable thumb for grasping branches, and this thumb survived among the prosimians and some New World monkeys. Second, either from it or independently, the truly opposable thumb appeared among the Old World monkeys and apes. Finally, the evolution of bipedalism freed the hand from all locomotor function and allowed the perfection of the precision grip by a lengthening of the thumb metacarpal and phalanx, a slight further modification of the wrist, and some development of the musculature of the thumb. By this time the fingers had straightened, and the familiar human hand evolved.

The hand has contributed as much as the eye to the making of *Homo sapiens*; together they gave us a new perception of our environment and, with the development of technology, increasing control over it.

Our Ancestors: Arboreal Climbers

It seems incontrovertible that our ancestors were climbing arboreal apes. Not only do we share many characteristics with the living climbing species, but *Australopithecus* carries an even greater variety of characteristics suggesting climbing. Oxnard has reviewed the evidence relating to the evolution of the human shoulder and concluded that the most likely origin for it lies in a fully arboreal ape genetically related to *Pan* but functionally similar to *Pongo*. Stern agrees, but adds that the ancestor was smaller than an orangutan, though it employed its long forelimb much as the living orangutan does. It used its hindlimb in a pronograde manner but was somewhat better adapted to this than the orangutan itself. Tuttle and Basmajian conclude that humans and *Pan* share a common heritage of arboreal adaptations including vertical climbing, hauling, hoisting, and suspensory behavior, while knuckle-walking played an inconsequential role in protohominid behavior. Fleagle and coworkers, in a summary paper have concluded that, on the basis of experimental work involving electromyographic studies of muscle activity and limb excursion in ceboids and apes, climbing and more suspensory modes such as brachiation entail similar patterns of muscle use and bone strain in the forelimb. It follows, then, that there is no longer any reason to attribute suspensory modes of locomotion such as brachiation to the ancestral hominid. Thus, with Andrews and Aiello, we can see the ancestral species as *forest-living climbers* of increasing body size. As they became heavier they increasingly engaged in below-branch feeding developing the short trunk and highly mobile forelimb of the living hominoids.

Humans do, of course, bear certain characteristics in the forelimbs that are adaptations to erect posture and that result from the downwardhanging position of the arms. Though we are a much more recent arrival on the earth than the kangaroo or the dinosaur, it is interesting that we share with them some reduction in our forearms. We can interpret this as a correlation of bipedalism and the development of manual skill.

Finally, we need not be surprised to find that the muscles that carry the weight of the body in climbers are, with their areas of attachment, reduced in humans. Although humans use the same muscles for carrying as the apes do for climbing, these muscles do much less work.

Though we bear the marks of apes in our bodies, we are profoundly different from any of them. As noted earlier, we actually share more of our locomotor morphology with *Alouatta* than with any living ape. This demonstrates with some force how far both we and the modern African apes have diverged from the condition of our shared ancestors.

8

Vegetarian Lifestyles in Life Science

Introduction

The number of insect species consuming green plants is slightly larger even than the number of species of green plants (at last count around 308,000). Insects aren't the only arthropods that have adopted a vegetarian lifestyle in a very big way; among the Arachnida, many species of mites are herbivorous as well.

One thing that has contributed to the diversification of herbivorous insects is that there are many ways to eat any particular plant. Leaf-chewers, like most grasshoppers and walkingsticks, simply tear off plant leaf tissue and swallow. Other insects with chewing mouthparts feed inside plant tissues. Leaf miners eat the layers of plant cells between the inner and outer layers of leaf cuticle. Some caterpillars, beetle grubs, and even maggots can make linear mines, which look like pale wavy lines on a leaf, or blotch mines, in which large leaf areas look white, since all the good green stuff has been eaten away. Many morphological features of leaf miners reflect their space limitations inside a leaf. Generally, they are very small and compressed or flattened. In these species, the mouthparts generally point forward instead of down, to give them a lower profile inside the leaf. Other herbivorous insects that lead a concealed existence bore into the stems of weeds or the trunks of trees or tunnel their way underground through the roots of plants.

Sucking insects generally feed on the vascular sap, or phloem, of plants, that moves from the leaves where photosynthesis takes place and which is therefore rich in carbohydrates, and on the xylem, which moves up from the roots carrying water and mineral nutrients from the soil. Sucking herbivorous insects are concentrated in the orders *Hemiptera* and *Homoptera* and include in their ranks (among many, many others) leaf bugs, stink bugs, chinch bugs, aphids, whiteflies, jumping plant lice, scale insects, treehoppers, leafhoppers, and cicadas. One notable accomplishment of these sucking insects (although a dubious one by human standards) is that they can manufacture two different types of saliva. One sort is injected into the plant and hardens to form a sheath in which the thin probing mouthparts can move around freely. Another kind of saliva, like more typical salivary secretions, breaks down starches and cell walls to make sucking easier.

Some insects can eat plant tissue by commandeering the plant's hormonal system in such a way that the plant is induced to produce bizarre and unusual growths, which provide the insect with a place to live and with nice nutritious tissue on which to feed. These feeding structures are called galls. Gallmaking insects can be found among the thrips, the aphids, beetles (notably weevils), moths, wasps, and flies; as well, there are some gallmaking noninsects, like mites (for that matter, galls can be induced by a whole host of other organisms, including viruses, bacteria, and nematodes). Although some plant species appear to tolerate galling insects with few ill effects, some gallmakers, like the Hessian fly, a serious pest of wheat, can greatly reduce yields or even outright kill the host plant. The scientific name of the Hessian fly*Mayetiola destructor* reflects its impact.

Some insects concentrate their plant feeding on the fruits or seeds of the plants. Seed feeders in particular can have a tremendous impact on plant populations because, by destroying a seed, a seed predator is actually destroying a whole potential plant (which would be quite a mouthful under other circumstances). Seeds are generally very nutritious, high in both fats and protein, because they contain just about everything a growing seedling will need. Seed feeders include in their ranks seed-sucking bugs as well as masticating caterpillars, beetle grubs, and even maggots and wasp grubs. Many seed predators can feed on seeds not only when they are growing on the plant but also when they fall to the ground, or when human beings harvest them and store them (or grind them up, make flour, bake cakes and cookies, and store *them*).

All plant-feeding insects require some basic equipment for adopting a plant diet. First, insects need sensory equipment for finding edible plant food. In some cases, this means finding anything green; polyphagous insects are capable of feeding on many different families of plants. Despite the fact that some of the world's most conspicuous plant pests are polyphagous (locusts and armyworms come to mind), about 90% of all herbivorous insects feed on three or fewer plant families; such species are said to be oligophagous. There are even some insects that can feed only on a single species of plant. The level of sophistication or complexity of the sensory systems employed for food-finding varies with the breadth of the diet. In general, herbivorous insects use olfactory receptors to detect volatile substances given off by the plants (such as essential oils). These receptors are often located on the antennae. Once on a plant surface, an insect can obtain additional information about its suitability as food by touching or tasting it. Insects may touch the plant with contact chemoreceptors located on the tarsi (legs) or even, in the case of females looking for a place to lay eggs, on the ovipositor. Finally, taste is the ultimate arbiter of acceptability. To taste plants, insects have mouthparts liberally equipped with gustatory chemoreceptors.

The next job of an insect *herbivore* is to consume the plant tissue, and, again, mouthparts are indispensible for this purpose. Insect mouthparts have undergone a remarkable number of changes to suit the different sorts of diets insects have adopted. For chewing plant tissue, the basic mouthpart plan has been modified only a little. These basic mouthparts can be clearly seen in the strictly vegetarian grasshoppers. The mandibles, the most anterior pair of segmented appendages associated with the head, do the bulk of the chewing. In most species, they are equipped with teeth or ridges to help with cutting, tearing, or shredding plant tissue.

For sucking plant fluids, the mouthparts are modified variously into devices that operate on the same principle as the drinking straw. In both *Hemiptera* and *Homoptera,* the mouthparts constitute a beak. The labium serves as a sheath for the remaining mouthparts, and the mandibles and maxillae are modified into piercing stylets. The two maxillae are grooved and fit together to form two channels, one for pumping out saliva and one for pumping in plant sap and the contents of broken plant cells. Lepidopterans consume nectar with sucking mouthparts built on a different plan. The proboscis of a butterfly is a long coiled tube made up of modified parts of the maxillae (the *galeae*). These are fused together to form a narrow channel. Like the *Homoptera* and *Hemiptera,* butterflies and moths have a set of muscles in their heads, the cibarial pump, which gives them the ability to suck up a fluid. Some plant-feeding thrips neither chew nor suck rather, they rasp away at plant tissue. To do so, they have a cone-shaped set of mouthparts with three stylets: the left mandible (the right one reduced to only a vestige), and modified parts of the two maxillae (the *laciniae*).

Once food is ingested, it has to be processed, and insects have a very adaptable alimentary system that has been variously modified for different kinds of plant diets. The most generalized system is that found in the grasshoppers (*Orthoptera*). The alimentary canal is partitioned into three general regions: the *foregut,* the midgut, and the *hindgut.* Valves or sphincters can close off each region to control the rate of passage of food. The foregut usually consists of a pharynx (right inside the mouth), an *esophagus,* a crop, and the *proventriculus.* In the crop, in some species, food is mingled with secretions from the salivary glands, which lie underneath the front part of the alimentary canal and empty into the mouth. The *proventriculus* in some species (e.g., in the *Orthoptera*) is armed with teeth and serves two functions: regulating the passage of food into the midgut and breaking it up into smaller pieces.

The midgut is the site where most food processing takes place. Unlike the hindgut and foregut, it is not lined with cuticle, so nutrient absorption can take place. Midguts are sites of active enzyme secretion. Digestive enzymes that break down fats, carbohydrates, and proteins are secreted into the midgut, and detoxification enzymes, which break down the poisons that are often found in plant food, are located in the cells lining the midgut. The midgut is lined with a peritrophic membrane, which allows dissolved nutrients to pass through but protects midgut digestive cells from damage by food particles and prevents the influx of microbes that may cause infection. It also serves as sort of a wrapper for waste products and is excreted with feces.

In the hindgut, water is absorbed and waste products are concentrated and packaged for excretion. At the front end of the hind gut are the Malpighian tubules. These take up water, salt, and waste products from the blood (hemolymph) and move them on through to the hindgut, where the water is resorbed and feces are formed. The number of Malpighian tubules varies with the species; while scale insects have only two and aphids have none, some grasshoppers have about 250. Wastes containing nitrogen are generally excreted in the form of uric acid, which is relatively nontoxic and which can be excreted dry without polluting the insect's immediate vicinity. Using uric acid as a waste product allows terrestrial insects to conserve water since other forms of nitrogenous wastes, such as ammonia, must be flushed out of the body in large volumes of water in order to avoid toxicity. The feces of most leaf-chewing herbivorous insects are very distinctive; they generally reflect the shape of the gut as they are extruded, like the way cookies are formed in a cookie press. There are actually keys to insect feces (called "frass" by those in the know), which can be used to identify the species that made the deposit.

The alimentary canal of sap-sucking species has been considerably modified in comparison to the basic body plan. Sap feeders consume a very dilute source of nutrients, particularly nitrogencontaining protein. Although phloem is rich in carbohydrates, it contains only about 1% nitrogen; xylem, only 0.05% nitrogen, is even more dilute. As a consequence, sap-feeders have to pass enormous volumes of fluid through their bodies to extract enough protein to survive. They are therefore specialized to eliminate large volumes of water very rapidly, so they don't end up diluting their internal organs. In the *Homoptera,* a section of the midgut is enlarged into a large bladder-like organ that lies right next to the front end of the hindgut and the Malpighian tubules; the whole arrangement is called a filter chamber. With the midgut lying along the full length of the hindgut, water can pass directly from one section of the gut to the other, for faster elimination. The excess water, along with its load of excess carbohydrate, passes out the anus onto the plant surface, to form a sticky sugary residue called honeydew.

In terms of absolute amounts, plants are the most abundant sort of food on earth. However, much plant material (the vast bulk, it seems) goes uneaten. Now, insects aren't keen on giving up on food for the taking, so it stands to reason that there must be something wrong with plants as a sole source of nutrition. Actually, for any animal, there are many basic problems associated with living on an exclusively plant food diet these problems are routinely faced by human vegetarians. One of the main problems with plant food is that it is relatively low in protein. At 0.05% nitrogen, xylem sap is barely nutritious enough to sustain life; heartwood and bark, at 0.3% nitrogen, aren't much better. Even relatively protein-rich foliage is rarely greater than 20% protein. Moreover, the protein in plant food is not of the right composition to suit the nutritional needs of most animals (that's why strict vegetarians have to eat a variety of different plants, so that the deficient proteins in each type of food balance or complement each other).

Water can also be in short supply for insect herbivores. Although many weedy species have relatively high water content, the leaves and seeds of woody species and wood itself can be critically low in water. Seeds and grains as well are not exactly juicy. On average, herbivorous insect bodies contain about 75% water, substantially higher than most of the plant food they consume, which means that many plant-feeding insects must take a drink of water every now and then (as many grasshoppers do) or they must carefully conserve the water that they take in with their food.

Plants are also loaded with indigestible substances. The glucose molecules that make up the carbohydrate cellulose, for example, are linked in such a way that the enzymes of most insects (and most other animals) can't break them apart (silverfish and a few of the higher termites are excepted). Also found in plants like grasses is the substance silica. Not only is this hard rocklike material (of which sand is made) indigestible, it also wears down even the relatively tough cuticular surfaces of insect mandibles that come in contact with it.

Plants tend to be relatively deficient in the vitamins and minerals that animals need to survive. Sodium, for example, is present at very low concentrations in plant tissue. All animals with nervous systems require significant amounts of dietary sodium to ensure proper nerve function. The low sodium content of plants is why conscientious farmers put up salt blocks or salt licks in their cow pastures for their cows. Some insects have "licked" the low sodium problem in a remarkable way. Butterflies of many species can often be seen gathering in so-called "puddle clubs" around mud puddles, bird droppings, and animal urine. It's believed that what attracts them to the puddles in the first place is the high level of sodium in these materials. Puddling in most species of *Lepidoptera* seems to be exclusively male, for reasons that are not totally understood; it may be that, when male butterflies mate and pass on a spermatophore, or sperm package, to females, they pack it full of sodium and thus require more sodium than do females to make it through their adult life. Since they can't get enough from nectar, the usual diet for a butterfly, they must go to alternative food sources, unattractive as they might appear at first glance.

One of the major obstacles that limits the activities of herbivorous insects is that all plants produce a number of chemical substances that appear to have no function other than to discourage insects and other would-be herbivores from consuming them. These secondary compounds (so named because they play no role in primary life processes of plants, such as respiration or photosynthesis) are extremely diverse; at least 100,000 different chemical structures have already been characterized from flowering plants. They can act in a variety of ways. Some are feeding deterrents that either prevent an insect from recognizing plant tissue as edible or render the plant tissue unpalatable. Other secondary compounds interfere with an insect's ability to extract nutrients from plant tissue and hinder growth and development. A tremendous number of plant chemicals are out-and-out poisons that can kill insects if consumed in sufficient quantity. The insecticidal nature of some plant chemicals has actually been exploited economically; the commercial insecticides nicotine and pyrethrin are both of botanical origin.

Despite all the formidable barriers to herbivory, though, just about every species of plant has its share of insect herbivores. Insects have a phenomenal capacity for overcoming the nutritional and chemical barriers that plants have erected to protect themselves. And once they overcome those barriers, they can eat a lot of plants. Even though individually insect herbivores are much smaller than, say, mammalian herbivores (like herds of buffalo), they still are capable of consuming amazing amounts of plant tissue. More plant material cycles through a population of grasshoppers, for example, than cycles through a population of meadow mice; the total energy flow through insect herbivores can be three to four times greater than the energy flow through mammalian herbivores. Insects can eat up to 75% (or more) of a given plant species in a particular spot during outbreak situations.

Not only do insects eat a lot of food, they also excrete a lot of waste, and they play an important role (out of proportion to their small size) in recycling nutrients in ecosystems. Under outbreak conditions, by defecating insect herbivores in forests account for 2 to 5% of the carbon, potassium, nitrogen, sodium, and calcium and over 8% of the phosphorus that eventually returns to the forest soil. Insect frass may thus represent a new source of plant fertilizer for the truly industrious entrepreneur.

Down on the farm

Fossil Evidence Suggests that insects first appeared on earth some 400 million years ago and it is likely that many of the early colonists of dry land were herbivorous. The angiosperms, or flowering plants, the dominant form of plant life on earth today, first appear in the fossil record about 100 million years ago, and humans, or *Homo sapiens,* only about 250,000 years ago. Humans, angiosperms, and herbivorous insects more or less peacefully cohabited the planet until about 10,000 years ago; at that time, the human perspective on the plant-eating habits of insects changed irrevocably. Concurrently on at least four continents, humans hit upon the idea that growing plants was a far more reliable way to obtain food than gathering plants. Thus, humans invented agriculture. Probably the first plants to be domesticated were the cereals corn in the Americas, rice in southeast Asia, wheat and barley in the Near East, and sorghum in Africa. Subsequently, additional plants were domesticated-legumes such as beans, peanuts, and soybeans; fruits such as coconut and banana; and roots and tubers, including beets, potatoes, sugarcane, yams, and cassava.

Nobody knows the precise details that led to the invention of agriculture, but it's generally thought that it was the result of an anonymous act of carelessness. During the Paleolithic period, or Old Stone Age, humans made their living by collecting seeds or grains from wild plants whenever they were available a sort of catch-as-catch-can existence necessarily nomadic because there was no single locality that produced enough food to support a human population year-round. It's likely that the whole world changed when somebody tossed some of the collected grain accidentally into a garbage pile. In the high-nutrient environment, the seeds germinated and some nameless genius saw the solution to the ageold solution of finding enough food to eat don't go out and search for it but grow your own. The invention of stone tools around 3000 B.C., which greatly facilitated working the soil as well as gathering the harvest, marked the beginning of the New Stone Age, or Neolithic period, and agriculture became firmly established and a viable enterprise.

One of the consequences of cultivating plants is that, for the first time in human existence, people were able to generate a surplus of food more than they could eat in a short period of time. This surplus could be stored in anticipation of the hard times ahead. Storage of food allowed people to abandon a nomadic lifestyle and to settle, in larger numbers, in one place. Surplusses of grain meant that domestic animals could be maintained yearround and a regular source of meat and dairy products improved people's diet (cholesterol notwithstanding). Soon towns and cities arose and civilization really began to take off. Art, science, culture, and eventually even university courses in entomology would develop, all as a result of someone tossing some grain in the garbage a long time ago.

In addition to affecting human lifestyles, the domestication and cultivation of plants greatly affected insect lifestyles. Selective breeding of plants, either deliberate or accidental, led to the establishment of agronomic traits that were highly desirable from a farmer's point of view. Dispersal mechanisms, for example, were bred out of plants, so seeds remained on the stem instead of falling to the ground and scattering. Growth and development became more synchronous, so harvesting activities could be concentrated into a relatively short time period. Edible seeds got larger and more numerous on each stem. Such changes were well-suited to insect herbivores as well, who, when feeding on cultivated plants, were faced (like farmers) with an abundant, predictable, highly localized food supply. In addition to altering the biology of the plants themselves, early agriculturalists also altered the ecology of the plants. Crop plants were irrigated and fertilized, so, from the insect perspective, water and nitrogen limitations on the nutritional suitability of plants were greatly reduced. Insect populations could increase to levels totally without precedent in the natural world much to the chagrin of hungry people, who for the first time in about 250,000 years on the planet had cause to resent the herbivory of insects.

This resentment grew all the more acute during and after the second Agricultural Revolution, which began around the fifteenth century and is continuing today. With the beginning of the Renaissance, world exploration introduced peoples of all countries to new and different crop plants. Plants were exchanged and established in places far distant from their places of origin. Sugarcane, for example, was indigenous to southeast Asia and was introduced extensively to North, Central, and South America; potatoes, native to South America, became staples in northern Europe. Thus, as cultures acquired new foods and fiber sources, plants acquired new insect enemies. Developments in science also created new markets for hitherto economically unimportant plants as sources of medicine and as sources of recreational drugs (including coffee, tea, and chocolate, as well as the more insidious tobacco and opium). Today, hundreds of plant species are producing products of economic importance in virtually every nation and as a consequence insect herbivory and its yield-reducing consequences have far greater economic importance than ever before.

Population Growth (*Locust Plagues and Insect Exorcisms*)

It is important to know something about the staggering ability that insects have to increase their numbers. Insects of the same species living in the same place at the same time constitute a population. Populations characteristically change in size over time; birth and immigration increase the size of a population while death and emigration decrease its size. The resulting average population size varies widely with species and locality. While there are some social species, like mound-building termites, that live in populations containing literally millions of individuals, there are, on the other hand, some populations of butterflies that contain only a few dozen individuals.

There are several aspects of insect reproduction that determine how fast populations can increase in size and how large populations can grow. One factor that can vary from population to population is how many offspring are produced each time an insect reproduces. A female German cockroach (*Blattella germanica*), for example, packs anywhere from thirty to forty-eight eggs into each ootheca, or egg capsule, that she produces. Cockroaches in an apartment with a careless tenant who leaves faucets dripping and crumbs all over the kitchen floor will be healthier and hardier and consequently can pack more eggs into an ootheca than the less fortunate cockroaches next door in an apartment occupied by a compulsively tidy tenant. Another important factor is how many times offspring are produced over the lifetime of an insect. German cockroaches produce on average about five oothecae during their reproductive life but more prolific individuals may produce twice that number. Finally (and not really obviously), the age when insects reach sexual maturity and begin to reproduce can have a big impact on the rate at which a population grows.

The age at first reproduction (or generation time, the time it takes to run through the generational cycle of egg, larva, pupa, adult, egg) doesn't seem at first to be so important in determining rates of population growth, but an example with more familiar vertebrates may make the concept more graspable. Mice reproduce when they're 18 days old, dogs and cats at 7 months or so, and humans not until they're at least 14 years old. Think how many litters a cat can have by the time it takes a person to even figure out what's going on. The actual record for individual cat reproduction is on the order of 420 kittens (produced by one Dusty of Texas between 1935 and 1952); in contrast, the record for humans was set in the eighteenth century by the first of the two wives of Theodore Vassilyev, who had sixty-nine children (including sixteen sets of twins, seven sets of triplets, and four sets of quadruplets). This is one record that not too many people are vying to break.

These mammalian records pale in comparison with insect potential for reproduction. Some fly species have a generation time of less than a week. Take *Drosophila melanogaster,* the humble fruit fly, for instance. Under ideal conditions, a generation can take a little less than two weeks, so up to twenty-five generations a year are possible. Every female can lay at least 100 eggs. So, even on the conservative side, starting with a single pair of fruit flies, assuming all offspring survive and all female offspring produce 100 eggs, in only one year there will be 1041 fruit flies. Packed tightly together at 1,000 per cubic inch, these flies would form a ball 96 million miles in diameter, a distance approximating that from the earth to the sun.

Entomologists love to make these sorts of calculations and in fact often get into disputes about them. Estimates on the number of offspring a single pair of house flies (*Musca domestica*) can produce in a year led to one such controversy. In 1906, L.O. Howard took a shot at estimating the potential number of offspring from a single house fly in one season in Washington, D.C. According to his calculations, if the fly began laying eggs in the beginning of the year, there would be 5,598,720,000 by September 15 enough, according to Howard, to cover the earth to a depth of 47 feet. These figures went unchallenged for over half a century until 1964, when Harold Oldroyd took exception and recalculated that "a layer of such thickness would cover only an area the size of Germany, but," as he observed, "that is still a lot of flies."

That we're not ankle-deep in house flies suggests that something out in nature is keeping populations in check. That something is the environment. Any given environment has only enough resources to support a limited number of individuals of one species indefinitely. This number is the carrying capacity of the environment. Population growth tends to increase until the carrying capacity of the environment is reached and then it either drops precipitously, levels off, or fluctuates cyclically around the carrying capacity.

There are two kinds of environmental factors that can cut down on population growth. One group of forces opposing unlimited growth acts independently of the size of the populations. Bad weather, for example, can wipe out everybody. A hurricane, tornado, or earthquake can bring about a huge toll in insect populations that is never reported on the nightly newscasts. Even less dramatic climatic vagaries can limit insect population growth. The temperature tolerance of insects varies with species. Generally there is an optimal range over which insects function most efficiently. Since insects are cold-blooded (heterothermic); their body temperatures are often dictated by the environment and sudden changes in temperature can prove deadly. The majority of insects cannot function in temperatures below freezing (32°F, 0°C). There are remarkable exceptions, of course. *Grylloblattids* (or rock crawlers), which live high atop mountains in the western U.S. and in eastern Asia, are most active at temperatures just around freezing. They are, however, not very tolerant of warm temperatures the heat of a human hand can kill them. In contrast, many species adapted to desert or thermal spring environments are extraordinarily tolerant of high temperatures; there are brine flies that live in hot springs in water temperatures exceeding 140°F.

Moisture is also a determinant of environmental suitability. As with temperature, most insects function most efficiently over a relatively narrow range of humidities. Again, there are exceptions-stored-product insects, for example, that can live at effectively zero percent humidity by extracting water from the foods they eat, and, at the other extreme, insects adapted to aquatic environments. Other determinants of climatic conditions wind, light, and soil and air chemistry, for example are also determinants of the range and population sizes of individual insect species.

Some forces, usually biological in origin, become more severe as population density increases. These so-called density-dependent factors include such sources of mortality as disease; the more crowded individuals are (even insects), the greater the chances for a sick individual to come into contact with and infect others. Like mammals, insects suffer from diseases caused by protozoa, bacteria, viruses, fungi, and nematodes. Other density-dependent mortality factors include competition for food. Plant-feeding insects compete for foliage or fruit to eat, and for places to lay their eggs; predaceous insects compete for prey. The greater the number of competitors, the more intense the competition. It's important to remember, however, that insects don't necessarily have to die in response to deteriorating conditions. One effect of higher population densities is that, when the carrying capacity of the environment is exceeded, the tough get going individuals can disperse to find greener, less crowded pastures, leaving behind them lower, more manageable population densities.

Possibly the most conspicuous example of the potential for insect population growth to get out of hand involves *Schistocerca gregaria,* the migratory locust. This species of grasshopper is native to the Middle East. Periodically, populations build up to enormous levels and swarms of individuals take off, a fantastic phenomenon called mass streamaway, in search of food to sustain themselves. Locust swarms can be up to 1,000 square kilometers (400 square miles) in extent and can contain up to 10 billion individuals. These swarms are maintained as cohesive units for days or weeks over thousands of miles.

Two factors are instrumental in leading to mass streamaway: crowding, which increases the rate of physical contact among growing nymphs, and inadequate nutrition during development. The complex transformation that leads to swarming in locusts was not well understood until the early part of the twentieth century, when detailed studies by Boris Uvarov revealed the mechanism.

Normally, the migratory locust takes the form called *Schistocerca gregaria var. solitaria.* These individuals lead solitary lives, getting together only for mating and egg-laying. However, when numbers increase in an area, nymphs are crowded and food becomes scarce. Under these conditions, a series of physiological changes takes place during the process of development. Nymphs begin to aggregate and form small groups that venture short distances from their place of birth. These treks increase in frequency with age. After five nymphal instars, when the nymph finally molts into the adult stage, it becomes clearly recognizable as the form called *Schistocerca gregaria var. gregaria.* This form differs in both color and morphology-the adults are pinkish, sexually undeveloped, and have relatively long wings for the length of their body. These individuals bask in the sun all day and begin to make short flights in apparently random directions as air temperatures rise. One day, when atmospheric conditions are just right, mass streamaway takes place; as if on signal, about two hours after sunrise one day, millions of individuals head off in the same direction within fifteen to twenty minutes of each other. As swarms pass overhead, other locusts join them, and eventually millions of locusts take to the air en masse.

Most swarms appear to move in a downwind direction. Locusts are not, however, just passively blown along for the ride even when wind speeds are less than 20 kilometers per hour (9 to 12 miles per hour), the typical air speed for a locust, locusts actively fly downwind. They can sustain these speeds for twenty hours or more, and individual swarms can move over 3,000 miles from their place of origin. Flying downwind, particularly in Africa and southwest Asia, is really a good move on the part of the locusts. Here's why: desert locusts invade some 11 million square miles of territory in that part of the world but breed in only about 5 million square miles. The area is seasonally dry and breeding is restricted to areas of seasonally high rainfall. Weather patterns are tied closely to the Intertropical Convergence Zone, a narrow zone 25 to 300 miles wide where northern- and southern-hemisphere air masses (the trade winds) converge. At convergence zones, air flows upward and, as it moves upward, it expands and cools. Because cool air can't hold as much moisture as warm air, this so-called adiabatic cooling results in heavy precipitation. Locust breeding grounds are concentrated in areas of convergent air flow. Swarms move downwind in spring from north to south; traveling downwind virtually guarantees deposition in the Intertropical Convergence Zone, where locusts can successfully concentrate for summer breeding. One reason for heading for this area of heavy rainfall is that locust egg cases are particularly prone to desiccation, so dry soil is totally unsuitable for producing children. Over her 6-month lifespan, a female can lay from 300 to 400 eggs in several batches.

Locusts are unwelcome visitors even when they travel in small numbers because of their voracious and undiscriminating appetites. To "eat like a horse" is an understatement a single locust can eat its own weight in food (0.10 ounce) every day, a feat few horses (at a thousand pounds or more) are capable of. Pound for pound, that's 60 to 100 times more than even a gluttonous human can eat in a day. Each square kilometer of locusts, amounting to 40 to 80 million individuals, can eat up to 250 metric tons of food daily enough to support a city of 80,000 people for the same period. In nonmetric terms, one square mile's worth of locusts 100 to 200 million individuals can consume 220 to 270 tons of food, or enough to feed 200,000 people.

Locusts have been literally plaguing humanity for centuries. Perhaps the most well-known literary allusion to locusts is in Exodus 10:3-6, 12-17, 19, the description of the eighth of the ten plagues of Egypt, inflicted upon the Egyptians when they would not grant freedom to their Israelite slaves:

> *And Moses and Aaron came in unto Pharaoh, and said unto him, thus saith the Lord God of the Hebrews tomorrow will I bring the locusts into thy coast: and they shall cover the face of the earth, that one cannot be able to see the earth: and they shall eat the residue of that which remaineth unto you from the hail, and shall eat every tree which groweth for you out of the field: and they shall fill thy houses and the houses of all thy servants, and the houses of all the Egyptians; which neither thy*

> *fathers, nor thy fathers' fathers have seen, since the day that they were upon the earth unto this day.*

The ninth plague, darkness, following on the heels of the eighth plague, is thought by some to result from the darkening of the sky by the enormous locusts swarms.

Throughout the Old Testament, there are frequent references to locusts, particularly when the idea of multitudes is to be conveyed (e.g., Jeremiah 46:23, "They shall cut down her forest, saith the Lord, though it cannot be searched; because they are more than the grasshoppers and are innumerable" or Judges 6:5, "For they came up with their cattle and their tents and they came as grasshoppers for multitude; for both they and their camels were without number.") So the ancient Hebrews were quite familiar with the ability of insect populations to increase to enormous sizes. They also had some awareness of locust lifecycles and biology. The Hebrew language has a number of different names for locusts, which may represent different developmental stages, sizes, or color phases.

The typical response to sudden visitation by locusts, before the advent of insecticides, was to seek help from divine powers. King Solomon, for example, issued a prayer of protection from locusts at the opening ceremony for the first temple (I Kings 8:37). It goes without saying that the Hebrews weren't the only ancient peoples with locust problems. The ancient Assyrians prayed to Ashur, a god always accompanied by an oversize locust companion, and wore locust talismans and amulets. The Greeks also had to deal with outbreaks of locusts and similarly resorted to divine intervention for assistance. The brass statue of Apollo Parnopios erected on the Acropolis at Athens was in recognition of his promise to drive away locusts (*parnopes* in Greek), which he evidently kept.

One safe thing about a promise to drive away locusts (or indeed any insect that suddenly undergoes a population explosion) is that in most cases it is easy to keep; eventually, density-dependent population regulation factors kick in and bring down the numbers in a given locality. So, no matter what people did to drive away locusts, it eventually appeared to be successful. It's not surprising, then, that prayers, spells, incantations, and denunciations continued to be regarded as effective anti-insect devices even down to contemporary times. Historically, Christians have subscribed to this belief. In the early days, court proceedings involving insects were strictly ecclesiastical because it was reasoned that insects were not answerable to civil authorities, rather only to divine authority. Thus, early on, it became the responsibility of the church to fathom the wishes of God or to foil the schemes of Satan (depending on who was responsible for the visitation in question). Excommunication was the ultimate weapon against insects, although there was a bit of confusion on this point; excommunication is effective only for church members and, since insects didn't pay taxes and weren't baptized, they technically couldn't claim church membership. Nonetheless, excommunications against insects could be obtained for a fee from Rome with great ease, although there was some controversy as to the advisability of such actions because they are God's creatures, to anathematize them was tantamount to criticizing His good judgment. Clever lawyers found their way around these formidable obstacles, however.

In one early case, at the close of the ninth century (A.D. 886), Rome and its surroundings were hit by a plague of locusts. Although peasants gathered and destroyed millions of locusts (a bounty was placed on their heads), the numbers continued to grow. In desperation, Pope Steven VI sprinkled the entire area with holy water, whereupon the locusts (according to contemporary accounts) disappeared.

Ecclesiastical explanations of outbreaks differ a bit from ecological explanations. According to church doctrine, as explained by a Jesuit priest, Father Bougeant, in 1739 in the form of a letter, entitled *Amusement Philosophique sur le Langage des Bestes*, to a female acquaintance, the death of any nonhuman organism sets free its devil, which is constrained to enter the egg or embryo of another animal to continue its "penal bondage to the flesh," deprived for eternity from a spiritual existence. This doctrine explains first of all why "all species of animals produce many more eggs or embryos than are necessary to propagate their kind and to provide for a normal increase." Thus, an outbreak of any particular species is the result of an excess of devils and a corresponding housing shortage.

This accounts for the prodigious clouds of locusts and countless hosts of caterpillars, which suddenly desolate our fields and gardens. The cause of these astonishing multiplications has been sought in cold, heat, rain and wind, but the real reason is that, at the time of their appearance, extraordinary quantities of animals have died or their embryos been destroyed, so that the devils that animated them were compelled to avail themselves at once of whatever species they found most ready to receive them, which would naturally be the superabundant eggs of insects.

Whether or not Father Bougeant was serious, many of his contemporaries were. An elaborate legal system was developed for the criminal prosecution and punishment of animals, including insects, by the Catholic Church. In the late fifteenth century, when a plague of locusts in Northern Italy was poised to devastate crops, the locusts were excommunicated. Recording this account, the Savoyan jurist Gaspard Bailly wrote:

The deadliest sword, wielded by our holy mother, the Church, to wit, the power of excommunication, which cutteth the dry wood and the green, sparing neither the quick nor the dead, and smiting not only rational beings, but turning its edge also against irrational creatures; since it hath been shown at sundry times and in divers places, that worms and insects, which were devouring the fruits of the earth, have been excommunicated and in obedience to the commands of the Church, have withdrawn from the cultivated fields to the places prescribed by the bishop who had been appointed to adjudge and to adjure them.

Ingenious explanations could be devised for the seeming lack of success of such excommunication. In 1338, locusts appeared in Botzen in the Tyrol, causing massive destruction to crops and fields. At the ecclesiastical court of Kaltern, about 10 miles south of the swarm, the parish priest was told to initiate proceedings against these locusts. The tribunal settled on a sentence of excommunication, which the parish priest executed. The failure of the locusts to disappear immediately was explained by the sins of the local people and their failure to keep up with paying their tithes.

Lest anyone think that the rights of locusts were trampled in these proceedings, there was an elaborate legal procedure to take them into account. First came the petition of the towns people seeking redress of grievances, followed by the official declaration of the townspeople. Then the plea for the defense of the insects was entered, followed by the response of the townsfolk, and the reply of the defendant. The decision of the bishop's proctor was followed by the sentence of the ecclesiastical judge. Locusts and other insects were duly appointed legal representation and many of the arguments raised on their behalf included the fact that they were owed some alternate place to make their living (being God's creatures); in some cases, the insects were ordered by the court to relocate to a specific piece of land, usually outside of town, designated for their exclusive use (although generally the townspeople retained water and mineral rights to the property). Other legal arguments in defense of the insects maintained that their presence in an area was not regarded as undesirable by all creatures, and birds on several occasions were called in as friendly witnesses by lawyers representing the insects.

Resorting to divine intervention to deal with a locust outbreak is hardly a vestige of the distant past. As recently as 1875, when the American Midwest was hard-hit by an outbreak of the Rocky Mountain locust *Melanoplus spretus,* the Governor of Missouriy proclaimed,

Wherefore, be it known that the 3rd day of June proximo is hereby appointed and set apart as a day of fasting and prayer, that the Almighty God may be invoked to remove from our midst those impending calamities, and to grant instead the blessings of abundance and plenty; and the people and all the officers of the States are hereby requested to desist, during that day, from their usual employments, and to assemble at their places of worship for humble and devout prayer, and to otherwise observe the day as one of fasting and prayer.

Granted, by the nineteenth century other, more scientific, measures were taken to combat the locust menace. Bounties were offered for collected eggs and nymphs (in Missouri, five dollars per bushel of eggs), and American ingenuity was put to new tests in designing devices to collect and destroy the adults (such devices included the King suction machine, a horse-drawn tank equipped with a fan, and the Adams locust-pan, a modified harvester that tossed locusts up into a pan full of oil).

By the start of the twentieth century, the development of insecticidal chemicals, along with a greater understanding of locust life history and reproduction, greatly improved the ability of humans to head off locust plagues for the first time in millennia. Poison baits, consisting of bran or other attractive food laced initially with arsenicals and subsequently with more potent insecticides, killed larger numbers of immature grasshoppers than was feasible by hand collecting (even with the assistance of a King suction machine). The use of aircraft to dispense insecticides also increased the mortality that humans could inflict on the locust multitudes. One aircraft, for example, was responsible for 400 tons of dead locusts in only fourteen minutes in Somalia in the early stages of swarm formation.

One of the hardest lessons to learn in combating locust swarms is that, because locusts pay no attention to political boundaries, effective control requires international cooperation. A series of outbreaks in Africa south of the Sahara in 1928 led to the First International Locust Conference in Rome in 1931. One outcome of that meeting was the establishment in London of the Anti-Locust Research Center, a facility dedicated to investigating the life (and death) of migratory locusts. In an example of the spirit of international cooperation, the British, deeply embroiled in World War II, sent several hundred troops into neutral Saudi Arabia in 1943 at the invitation of King Abd al'Aziz al Sa'ud to assist in an international locust eradication effort. Although the focus of the war had by then shifted away from the Middle East, the British invested troops, time, and money into heading off the locust plague to protect the civilian populations there and to free Allied ships for transport of military troops and supplies to the Pacific, rather than for transport of humanitarian aid to the Middle East.

Today, locust surveillance is carried out not only by European and American satellites but also by African nomads; spray programs are coordinated by the United Nations Food and Agriculture Organization's Emergency Center for Locust Operations in Rome. By preemptive strikes and preventative maintenance, the world has been free of major plagues of red locust and African migratory locusts for almost forty years, although desert locusts, with a much broader range, continue to wreak havoc with crop production throughout the sixty-five countries in which they breed.

Useful Homopterans (Working for Scale)

Members of the insect order Homoptera, as strict plant-feeders, are economically undesirable associates of humans. Taxonomists divide the order into two suborders. Members of the *Auchenorrhyncha* are generally active and mobile and include the cicadas, treehoppers, froghoppers, and leafhoppers. In contrast, species of *Sternorrhyncha* are inactive or sedentary. Aphids, whiteflies, and scale insects all spend much of their lives within a few feet of where they first began to feed as crawlers, or newly hatched nymphs.

There are species in both suborders that are serious economic pests of cultivated plants, reducing yields by removing large quantities of sap or causing mortality by mechanically blocking the vascular system and interfering with nutrient transport. Some of these species compound the damage they do by acting as vectors of plant diseases. In fact, over 90% of plant viruses are transmitted by homopterans. Leafhoppers can serve as vectors of a variety of viral pathogens of crop plants and are involved in the transmission of such diseases as alfalfa dwarf, alfalfa witch's broom, aster yellows, clover phyllody, clover stunt, corn stunt, elm phloem necrosis, Pierce's disease of grapes, peach X disease, phony peach disease, potato yellow dwarf, sugarbeet Argentine curly top, sugarbeet yellow wilt, and tomato-Brazilian curly top. Aphids, on the other hand, are vectors of sugarcane mosaic, parsnip yellow fleck, carrot redleaf, cauliflower mosaic, cabbage black ringspot, and a number of other plant viruses.

Plant viruses can be transmitted by homopterans in two ways. Stylet-borne viruses are carried mechanically from one plant to another on the long piercing mouthparts of the insect. Circulative viruses, in contrast, remain with the insect for a considerable length of time often through the entire life cycle. Some circulative viruses (especially those transmitted by leafhoppers) actually reproduce inside the body of the insect vector.

Although in general homopterans are considered economic pests, several species have proved to be economic boons. These species are almost exclusively members of the *Sternorrhyncha.* As sedentary as they are, they would be vulnerable to predators but for the fact that they can produce various forms of body armor: waxes, fluffs, powders, filaments, or hard plates. Mealybugs owe their name to the farinaceous white "meal" that covers the soft vulnerable bodies of all life stages except the first instar nymph and the winged male. Armored scales are protected by a hard resinous coat secreted by glands and permanently attached to the body. One proof that human ingenuity knows no bounds is the fact that over the centuries people have devised uses for the various fluffs, powders, and waxes that homopterans developed to keep themselves safe from harm to the point that some of the economically more profitable species have actually been "domesticated."

One of the more spectacular success stories involves lac and its derivatives. Lac is the waxy platelike secretion of a small scale insect called, appropriately enough, *Laccifer lacca,* in the family *Lacciferidae*. The secretion is the only animal resin of any commercial value, and it is chemically a complex mixture of various polyhydroxy acids. The word "lac" derives from the Hindi word meaning "hundred thousand" and refers to the huge aggregations that these insects form on their host trees (an etymological aside: the word "lox," referring to smoked salmon, is derived from the same root and describes the population densities of spawning salmon). Host species include fig, banyan, and a number of other species in Indochina, the Philippines, and Sri Lanka. As far as their biology goes, nymphs of both sexes crawl out from under the laccovered shell of their mother and quickly plug into whatever food source they can find. In densities of about 150 per square inch, they sit and suck, secreting lac from almost all over their bodies. As they suck, they go through three molts, progressively losing seemingly vital parts of their bodies, including among other things, legs, eyes, and antennae. For the females things go on pretty much as they did before, but males undergo a pseudopupation stage in which they develop new legs, eyes, antennae, and two wings. They then emerge and go out looking for sex. They really don't have too far to look the ratio of males to females in most lac bug populations is somewhere around 1:5,000, and the females, which produce sex pheromones to guide their prospective mates' search, are absolutely immobilized inside their little resinous cases, so the challenge to the male is really minimal. In any case, after mating, the males die and cease to be of interest to anybody.

The females, on the other hand, are of great economic interest and importance. They produce an orange-colored resinous secretion, or stick-lac, which is simply the coalesced bodies of breeding females. Huge populations can coat tree limbs with resin up to onehalf inch in thickness. This stick-lac is ground, washed, and filtered through muslin bags; the filtrate is passed over water-cooled drums to form flat flakes, the "shell-lac" of commerce. Alternatively, seed or grain lac can be ground by hand in stone mills or stuffed into long thin bags, heated over a fire, scraped, basted with water, spread onto a porcelain cylinder, and pressed flat and stretched wide with hands, feet, and teeth. Shellac color ranges from "ruby" to "blonde"; orange-brown is the purest form and white shellac is obtained by bleaching with chlorine.

Most of the world's lac is produced in India, which makes about 40 million pounds annually. It's an important source of subsidiary income to many of the poorer districts in Bihar, Orissa, Assam, and several other states. Lac's uses, even in today's synthetic society, are many and varied due in part to its physical properties, which include thermoplasticity, film-forming capacity, high resistance to organic solvents, excellent adhesion to a variety of surfaces, and good electrical properties. Moreover, unlike synthetic resins, it's completely biodegradable, nonpoisonous, and odorless. Lac has been used in insulators, electrical apparatus, and sealants; since it dries dust-free in ten minutes, it is invaluable in shoe and floor polishes and printing inks. Chippendale, Hepplewhite, and Sheraton, all famous furniture makers, used nothing else; championship bowling alleys are usually shellacked before an official match. Shellac can be found in playing card gloss, chocolate glazes, jewelry settings, billiard balls, mirror backing, and ironically enough, insecticide sprays (where it seals in the active ingredients).

Historically, the most important use of lac was in the record industry. The first shellac-based record was invented in 1907 by T.H. MacDonald and was a rather primitive affair consisting of two shellac-coated pieces of paper sandwiched together. World War I made shellac rather difficult to obtain, so there were a few abortive attempts to produce a synthetic replacement. None were very successful, so after the war shellac became very important for the growing record industry. In 1927-1928, Great Britain, Germany, and France collectively produced 260 million records, representing 18,000 tons of shellac. World War II, though, fixed it for the shellac industry. The unavailability of the product encouraged American scientists to come up with synthetic substitutes (such as vinyl), which proved to be far superior to the real thing, since shellac records are bulky, breakable and subject to extraneous surface noises (although, ironically, today vinyl records barely hold their own against compact discs for the very same reasons). About the only place in the world today where shellac is used at all in record making is, not surprisingly, India. Lac remains an important commodity in Thailand as well, and in China, where lac production was reinvigorated by government investment in 1962. Not only is lac resin important in the production of varnish, but the fire-resistant host trees of the lac scale are important in maintaining the integrity of swidden (slash and burn) agroecosystems.

Scale insects have commercial value in art and artifacts not only as a source of varnish but also as a source of dye. *Dactylopius coccus*, the cochineal scale, is not much more impressive at first sight than is the lac scale, but at least it has legs and eyes. The cochineal scale confines its feeding to cactus, specifically to species of prickly pear (*Opuntia*, particularly *O. cochinillifera*) which grow in Mexico and parts of the southwestern United States. Aztecs in Mexico and Pima Indians in Arizona cultivated them and used them in the manufacture of dyestuffs. The product was so valuable to them that cochineal was demanded as a medium of tribute from conquered peoples. Hernando Cortez was the first Westerner to report its use, in 1518, on his visit to Mexico. In 1523, Cortez received orders from the Spanish ministry to "multiply the commodity" and within a few years the dye was a major article of European commerce.

It wasn't for another 150 years, however, that people realized that cochineal was an insect. Anton von Leeuwenhoek and his newly invented microscope were needed to establish the fact. A man's entire fortune was lost in 1725 when he wagered that cochineal was not an insect; the story was published by the jubilant winner of the bet (in Dutch, French, and Spanish, no less) in a book titled *The History of cochineal proved by authentic Documents*. Spain attempted to monopolize the supply and forbade exportation of live insects. Several attempts by French and Dutch adventurers to smuggle the stuff out met with failure. It wasn't until 1833 that the French successfully introduced cochineal cultivation into French Algeria. Spain, faced with continental competition, then established a cochineal industry in the Canary Islands. In Europe, cochineal was used extensively in cosmetics, specifically rouge and fingernail polish. It found further use in medicine and fabric dyeing until, at the turn of the twentieth century, it was largely replaced by synthetic aniline dyes. Cochineal had just about disappeared from the economic forefront when interest in natural dyes was rekindled as a result of the discovery that many of the more popular synthetic dyes have carcinogenic properties (a finding that quickly led to a ban on their use in many foodstuffs). Cochineal is used occasionally (particularly in Europe) in medicine and food; Peru and the Canary Islands are the principal producers. Its continued use in cosmetics has caused some interesting cultural problems; lipsticks made with cochineal are not considered kosher by Orthodox Jews, who have strict rules about the use of animal products, even those that are not eaten.

The life cycle of the cochineal scale is fairly typical, as the lives of scale insects go. Females form small groups (from four to ten) on the nopal cactus and secrete a white cottony wax material (which protects against desiccation in the dry desert climate). After a few weeks of sucking cactus fluids, the female becomes sexually mature, although in the process she does not undergo any conspicuous changes in morphology other than enlarging to the size of a small pea (this sort of metamorphosis, in which immature morphology persists into adulthood, is known as neoteny). The male, in contrast, at its final molt acquires wings, flies to find females, mates with them, and dies (all in about a week, since he has no mouthparts with which to feed). The fertilized eggs hatch quickly into nymphs. The females are especially prized for dye production-they come equipped with a maroon pigment throughout their bodies, identified in 1910 as carminic acid. It amounts to about 10% of the total body weight of the female scale insect. This pigment may serve not only as a repellent to would-be predators such as ants it tastes repulsively bitter but also as an advertisement of the unsuitability of the cochineal as a menu item.

To produce the dye for commercial use, the cochineal are scraped off their cactus leaves, plunged into boiling water to kill them and to remove their wax coating, dried in the sun, and ground into a fine powder. About 25,000 insects are needed to produce about a pound of cochineal, and about three pounds of powdered insect make a pound of dye. Material to be dyed is boiled in a water extract of the powder along with leaves of a plant called tejute, which contains oxalic acid, a substance that is an intensifier and mordant (a substance that can combine with a dye to make an insoluble compound fixed in the fiber of fabric) for carminic acid. In pre-Columbian days, one of the more commonly used mordants was human urine; the number of chamber pots unearthed in the Aztec capitol was puzzling to archaeologists until the procedures for producing cochineal dyes were discovered.

Yet another scale insect has brought color into the drab lives of humans for centuries. Use of kermes dye, from *Kermococcus vermilius,* antedates Moses; in fact it probably was the source for one of the three colors specified for use in the curtains of the Tabernacle and for the robes of Aaron. Evidence exists for the use of kermes by the ancient Sumerians. The name "kermes" is mysterious in derivation (it seems to come from the Sanskrit meaning "worm") but it is fairly certain that the insects were native to the Mediterranean and Near East, where they feed on species of oak. The Arabs, by the way, are the ones who called the insect *kermes.* To Dioscorides, the Greek, the insect was known as *coccus.* In the Middle Ages, the name *vermiculus,* or little worm, came into vogue. To some extent, the dye has left its mark on our own culture through its linguistic reincarnations. From *kermes* comes the words "carmine" and "crimson," and from *vermiculus* comes "vermilion." The word "scarlet" in fact is derived from yet another name for this species *sikarlat,* a Persian term meaning "red" and latinized to *scarlatum.* At one time, kermes was the reddest red color known and was used in many of the medieval Flemish tapestries. In Venice, dye production was strictly regulated. Adulteration of kermes with madder by dyers in the city of Lucca was punishable by a fine of 100 lira or loss of the offender's right hand. The Red Coats of British soldiers and the red fez characteristic of the Turks (and later Shriners) owed their brilliant red color to kermes.

The females, globular in shape and about 7 millimeters in diameter (less than 1/3 inch) at maturity, can produce up to 3,000 eggs. Since there's only one generation annually, production is necessarily limited. Moreover, harvesting kermes is rather labor-intensive-each insect is individually scraped off its twig by fingernail. On a good day, a person can harvest a little over two pounds of scales, which will lose 70% of their body weight in the drying process to yield only a little over half a pound of dry scale for a hard day's work. The principal color component of kermes dye is an anthraquinone pigment, kermesic acid, similar in structure to carminic acid, the principal color component of cochineal.

Scale insects have historically been important as a source of food as well as a source of dyes or pigments. A species familiar to desert travelers in the Middle East was *Trabutina mannipara,* a species that feeds on the twigs of tamarisk bushes growing in the central Sinai mountains. This insect was treasured not for its color but for its excrement the sweet honeydew produced as a result of the rather inefficient method homopterans use for extracting nutrition from plant sap. Several lines of evidence suggest that the manna that rained down from heaven to sustain the Israelites during their ordeal in the desert was actually the honeydew of *T. mannipara.*

Location is one thing beginning in Elim and ending at Rephidim, the biblical appearance of manna is consistent with the northern limit of distribution of the manna scale today. Timing is another suggestive factor manna was discovered on the fifteenth day of the second month after the Exodus (late May or early June), exactly when honeydew production commences in contemporary times. The Bible also states that manna fell from heaven at night, a time when honeydew is likely to accumulate, since ants and other insects fond of sweetness are not actively harvesting it. Manna was said to resemble cumin seeds in appearance and bdellion resin in its sticky texture, an accurate description of the honeydew on both counts. Finally, the taste of manna is consistent with its identity as honeydew, since there are very, very few other sweet things that one is likely to run into when wandering in the desert.

Dyes, varnishes, and desert snacks notwithstanding, the greatest economic contribution made by scale insects is, in all probability, weed control. Biological control is the term for the use of natural enemies to eliminate an undesirable species. If the undesirable species targeted for elimination is a weed, then insects are generally the control agents of choice. Employing biological control is simply exploiting the fact that many herbivorous insects have a devastating impact on the plants on which they feed. Many species of scale insects have been deliberately introduced into areas where weeds have run rampant. They're ideal for this purpose they are often highly oligophagous and, passionately devoted to their host plants, do not pose a threat to any other possibly valuable plant species in areas of introduction. Species of cochineal have been successfully introduced into India, Sri Lanka, South Africa, Australia, Mauritius, New Caledonia, Celebes, Java, Hawaii, East Africa, the Leeward Islands, and even the United States in order to control prickly pear cactus (*Opuntia vulgaris* and other species). The nymphs and adult females inflict mortality on their prickly pear hosts due to the fact that, while they are withdrawing sap, they are at the same time injecting a toxic substance that can cause premature fruit drop, loss of pads, chlorosis (loss of chlorophyll), necrosis (tissue death), and eventual collapse of the plant.

The successes of weed biocontrol by homopterans and other herbivorous insects have often been spectacular and any of these success stories could serve to illustrate the fact that insects really can regulate plant populations. Probably Klamath weed control provides the best example. *Hypericum perforatum* was called Klamath weed because it was first discovered in the United States along the Klamath River in California in 1900 (it's otherwise known as St. Johnswort). By 1944 there were 2 million acres of Klamath weed choking off quality rangeland. Cattle and sheep died after eating it, due to the presence of a light-activated toxin, hypericin, in the foliage, and land prices dropped to less than one-third of uninfested neighboring areas. In Australia, the same plant was being devastated by a small leaf beetle called *Chrysolina quadrigemina.*

Shipments of the beetle from Australia commenced in 1944 (the beetle is actually native to Europe, but World War II prevented direct transport across the Atlantic). The first problem with settling the beetles in their new home was that the beetles were seasonally reversed and had to be acclimated to the Northern hemisphere summers from June to September. By 1946, after numerous tests on crop plants to ensure that, as reported, *C. quadrigemina* accepts only Klamath weed as a host plant, five thousand beetles were released. After three generations, there were thousands of beetles and by 1950 there were millions. Beetle grubs feed on overwintering plants and prevent them from flowering and fruiting in subsequent years. After a few years, there are no more root reserves and the plants die. By 1956, Klamath weed was a dim memory throughout much of California. The plant persists in shady areas only (beetles won't fly into shade to oviposit). From 1953 to 1959, increases in land values and savings in herbicides amounted to 9 million dollars and cattle grazing on the reclaimed land were worth $2 million per year, an impressive return on an investment amounting to no more than a couple of thousand dollars.

Gallmakers ("gall enough in thy ink")

A gall is an area of abnormal plant growth produced in response to a "foreign agent," including, among other things, plant-feeding insects. Galls can occur on just about any part of a plant from root tips to flowers and seeds. On leaves alone there are galls on stipules, petioles, veins, and leaf blades. They are often recognized (and classified) by their appearance; on leaf blades alone there are Filzgalls, fleshy galls, kammergalls, fold galls, roll galls, roll and fold galls, pit galls, pocket galls, oak-apple galls, and lenticular galls, to name but a few.

For thousands of years, the origin of galls was quite a mystery; Theophrastus, the "father of botany," remarked over two thousand years ago that most of the structures seemed to contain insects inside them but he thought that the appearance of these cecidozoans, or gall inhabitants, was strictly coincidental. Even the great Francesco Redi, who in the seventeenth century disproved that maggots appear spontaneously in rotting meat, was sufficiently baffled by the appearance of insects inside galls as to maintain a strict silence on the subject in his publications.

In point of fact, the reason that there is so often an insect inside a gall is that the insect is at least partly responsible for gall formation in the first place. This fact was experimentally confirmed in the seventeenth century by both Martin Lister in England (nephew of the physician to Queen Anne) and Marcello Malpighi in Italy (and described in his book *De Gallis*). Gall formation begins when a female lays an egg inside the plant; in response, the plant tissues begin to grow abnormally. Plant cells enlarge and multiply wildly. Insect galls are generally built on a similar plan. Lining the central chamber, where the insect makes its home, is a layer of protein-rich nutritive tissue; the cells in this layer often have multiple nuclei and are physiologically active, synthesizing carbohydrates and proteins. Immediately outside the central chamber is a layer of hard, thickwalled cells. These cells are rich in indigestible substances, such as lignin and tannin. This layer has been called the "protective layer," although its function is still not really certain. Outside this layer are layers of normal plant parenchyma and epidermal cells.

For a long time, no one could agree on whether plants produced galls to protect themselves from further damage by an insect enemy to seal it off and prevent its access to other plant tissues-or whether insects stimulate a plant to produce a gall in order to secure a safe haven and easy meals. As is so often the case in scientific controversies, it turns out that gall development depends on both the insect and the plant. When a female inserts an egg into plant tissue, she also injects a few other things along with it; some of these substances may have hormonal effects on plants and alter growth patterns. The suggestion has even been made that the insect injects a mutualistic virus into the plant, which can insinuate itself into the plant's genetic material and take over regulating the genes that control growth. While the disruption of growth patterns is the most conspicuous consequence of gall insect infestation, plant physiology is substantially altered as well. Generally, the production of nutritive substances such as carbohydrates is greatly increased. In fact, the galls of certain species of mites, aphids, wasps, and flies actively secrete a sugary secretion or honeydew. Such a secretion may attract ants and other predators that could protect the gall inhabitant from potential enemies.

The interaction between a gall-forming species and its host is unique; even if many different species of gall-forming insects infest the same host plant, each insect species produces a unique, characteristic gall morphology. Not all insects that lay eggs in plants can induce gall formation. Among the talented few are aphids, caterpillars, beetle grubs, thrips, wasps, and maggots; in addition there are a couple of families of mites that can cause galls to form on plants. The lion's share of gall-formers is found in only two families: the Cecidomyiidae, a family of gall flies, and the Cynipidae, a family of tiny gall wasps. In North America, these two families make up 70% of about 1,700 species of gallmaking insects. The different families of gallmaking insects are not at all alike in their plant preferences; while the cynipid wasps, for example, concentrate their efforts on oaks, the gallmaking aphids attack a tremendous variety of plants. In addition, species that induce gall formation vary in their ability to infest their hosts while there are rarely more than one or two goldenrod ball gall flies on a single plant, each housing only one maggot, there can be as many as half a million cynipid wasp galls on a single oak tree.

By the same token, not every plant responds to the presence of an insect egg in its vital parts by forming a gall. Gall formation is relatively rare among the ferns, mosses, conifers, and even the grasses; the vast majority (over 90%) of species that do form galls are flowering plants. Although galls have been found in at least half the families of flowering plants, the families hardest-hit by gallmaking insects are the Rosaceae, or rose family; the Compositae, or daisy family; and the Fagaceae, or oak family. In North America, the species in the daisy family make up about one-sixth of all plants susceptible to gallmaking insects.

Gallmaking is a specialized lifestyle to which not all insects are suited. Insects that develop inside galls have had to make a number of adjustments. A problem faced by gallmakers is relatively unusual in the insect world. Since gallmakers are confined in a small space, from which they cannot escape until development is complete, there is a premium on hygiene and sanitation they can't afford to contaminate their living quarters. As a result, many species of gallmakers have no functional anus. All of the waste products accumulated over the course of development are stored in the body and this collected waste material, or meconium, is voided only when the gallmaker metamorphoses into an adult.

On the other hand, unlike most plant-feeding insects, the gallinhabiting stages of gallmakers have no need to seek out food - they're literally surrounded by it. As a result, gall inhabitants often lack legs or have reduced numbers of legs; gallmaking mites, in the family Eriophyiidae, have only two pairs of legs instead of the typical arachnid four pairs. Also frequently absent are antennae, and with the antennae the sensory equipment they would use for locating host plants at a distance. The more mobile adult stages of these species are generally fully equipped with wings, legs, and sensory apparatus, since it is their responsibility not only to find host plants for their offspring but to find potential mates as well.

From the perspective of an insect gallmaker, living in a gall is a mixed blessing. The lignified outer wall prevents entry by many types of enemies. Gallmaking sawflies, for example, are free from attack by parasitic flies, who lack ovipositors long or strong enough to penetrate the protective layer. Unlike their close relatives that feed externally on leaf tissue and pupate in cocoons in the soil, gallmaking sawflies are also free from parasitism in the cocoon stage by wasps.

However, the protective layer that keeps predators out can also prevent the gallmaker from escaping in case a predator does manage to gain access to the central chamber. Mortality of gallmakers can be very high as a result. While most gall midges are proper gallmakers, a few rogue species eat fellow gall midges that do make galls. Not only are there interloping predators, there is also an assortment of inquilines, or "uninvited guests" other plant-feeding species attracted to the rich layer of nutritive tissue lining the central chamber. Aside from the fact that these inquilines rob gallmakers of their food, many of these species are not averse to consuming the gall occupant along with the plant tissue.

One final problem faced by gallmakers is escaping from the gall after the final molt to adulthood. Whereas mouthparts of immature stages of gallmakers are ideally suited for tunneling through plant tissue, the mouthparts of the adult stages of many gallmakers (particularly moths and flies) are woefully inadequate for this task. As a result, mature larvae often prepare escape tunnels immediately before they pupate, excavating a path to the outside and scraping away at plant epidermis until only a weak "window" remains behind. Upon completion of development, the adult can simply push its way through the window to the outside world.

Plants differ in the extent to which they suffer from attack by gallmakers. Despite the fact that at least six species of jumping plant lice can form hundreds of galls on hackberry trees, the trees never seem to suffer any ill effects. In contrast, a single goldenrod ball gall fly can reduce the production of seeds by a goldenrod by almost half. Gallmakers that attack crop plants, like the alfalfa gall midge *Asphondylia websteri,* can significantly reduce yields and cause economic damage.

In one case, the ability of a gallmaking insect to kill an economically important plant had world-shaking consequences. The wine industry of France was dealt a devastating blow in 1860, when *Phylloxera (Daktulosphaira) vitifoliae,* the grape phylloxera, was accidentally introduced from North America an event regarded as a viticultural disaster of unparalleled magnitude. The grape phylloxera is a tiny, wingless aphid, less than 1/20 inch long, that forms galls on both the leaves and roots of grapevines. While the leaf galls rarely cause much damage, the root galls can stunt or kill vines in three to ten years. Within only twenty-five years, over 2 1/2 million acres of grapevines (amounting to fully one-third of France's wineproducing grapes) had been ruined.

That France has maintained its reputation for producing fine wines is the result of careful entomological observation and great ingenuity on the part of French enologists. Noticing that the native North American fox grape *Vitis labrusca* did not readily succumb to grape phylloxera, the French imported rootstocks of this resistant species; in France, stems of the French wine grape *Vitis vinifera* were grafted onto the hardy American rootstocks. The rescue mission cost the French nation some 10 billion francs but, by restoring the French wine industry to prominence, it saved that nation untold billions and allowed the world to continue, with but a brief interruption, to enjoy fine French vintages throughout the twentieth century.

American wine, however, has not fared as well of late. In 1986, a new and virulent form of phylloxera made an appearance in California vineyards. This particular phylloxera reproduces only asexually and feeds exclusively on roots. By 1993, over 10,000 acres in Sonoma and Napa counties (premier wine country) were infested. Failure to control phylloxera either by pesticide applications or development of resistant rootstocks has led at least one enterprising vinter to market a rosé made out of infested Cabernet Sauvignon vines called "Bug Creek" wine, with 10% of the proceeds from sales of the rosé earmarked for research sponsored by the American Vineyard Foundation to stamp out phylloxera.

Gall insects aren't entirely bad news. Their ability to kill certain plants has been put to use in a number of weed biocontrol programs. Among the notable successes was the introduction of *Procecidochares utilis*, the pamakani gall fly, into Hawaii to control Mexican devil weed, *Eupatorium adenophorum*, an invasive weed of pastureland that was accidentally introduced into Maui in the midnineteenth century. The products of gallmakers, as well as their plant-consuming services, have been of economic importance for centuries. Galls, particularly those on certain species of trees, are very rich in tannic acid, a mixture of compounds, including gallic and ellagic acids, esterified to glucose. When the structure of tannic acid was determined and its component parts identified, they were named in honor of their source gallic acid and ellagic acid (from the French word *galle* spelled backwards). The Turkish oak gallnut is about 50 to 60% tannic acid and the Chinese gallnut amost 70%.

Tannic acid has several rather distinctive properties. Among other things, it is highly toxic to microbes. This may be the reason that gallmakers so often induce tannin production by their hosts-high tannin levels may protect gall insects from fungal or bacterial invasion. Yet another distinctive property is its tendency, when combined with iron salts, to form a blue-black precipitate. This is the substance that has, since about 2000 B.C., been used as ink. Ink made from galls over the centuries has been considered highly desirable. Dioscorides published a recipe for a gall-based ink in A.D. 50. Ink made from "oak apples" was used extensively by medieval monks in the ninth and tenth centuries for manuscript illumination. In some cases, certain types of documents could by law be recorded only with a gall-based ink. The Aleppo gall, produced on oaks in eastern Europe and Western Asia by the cynipid wasp *Cynips gallae-tinctoriae,* was particularly prized for its quality and was "specified in formulas for inks used by the United States Treasury, the Bank of England, the German Chancellery, and the Danish government". As recently as 1891, a report from the Massachusetts Record Commission on Record Inks and Paper concluded that ink from gallnuts was permanent and, if faded, even restorable by application of a solution of tannic acid.

Galls have also been useful in the dyeing of leather and of other materials. Again, the Aleppo gall has over the centuries been recommended for producing permanent black dye. As recently as 1914, over $17,000 worth of gallnuts were imported from Baghdad for the purpose of dyeing sealskin furs. Another gall, produced on oaks (such as *Quercus infectoria*), has been used in the production of a dye known as Turkey red. The gall, formed by the wasp *Cynips insana,* is variously known as the "mad apple of Sodom, Dead Sea Fruit, or Mecca or Bussorah gall". It is used locally in Asia Minor (in Basra). Because of their ability to produce a permanent pigment or dye, galls have a long historical association with cosmetic use. Over 2,000 years ago, Pliny the Elder wrote that the Aleppo gall was widely used not only for dyeing furs or leather but for dyeing hair black.

Due to their tannic acid content, galls also have been used in the process of tanning leather. The tanning process is one in which proteins in animal hides are complexed and rendered impervious to rot it's the fundamental process of making leather. Tannic acid is an excellent binder of proteins; historical records of the use of galls for tanning purposes date back to the first century. More recently, the binding properties of tannic acid have been exploited to the detriment of plant-feeding insects in general; nicotine and other organic insecticides have historically been complexed with tannic acid to stabilize them for longer shelf life.

While gall products such as tannic acid have been important throughout the world, galls themselves have been of value to people as well. The most direct use of galls for ornamentation is certainly the use to which the Aguaruna of Peru put galls found on a tree they call duship (*Licania cecidiophora*). The Aguaruna, who live in rain forests and farm manioc, collect the gall-infested leaves of the duship, which fall from the tree twice a year, and remove the galls in order to use them to string, like beads, in necklaces. These necklaces can contain as many as forty looped strands and contain as many as 41,600 galls. While not as durable as gemstones, galls are remarkably sturdy as far as plant products are concerned; duship capes collected half a century ago are still in good shape, possibly because the high tannin content of the galls protects them from decay.

Silk and Silkworms

Silk in its many guises (satin, shantung, peau de soie, brocade, organdy, taffeta, or chiffon) a fabric of elegance and expense, is basically nothing more than insect saliva. More precisely, silk is the threadlike viscous secretion of the salivary glands of a caterpillar called *Bombyx mori*, a member of the moth family Bombycidae. Now all caterpillars make silk, but few in as grand a style as the silkworm caterpillar. The salivary glands, stretched to their full length, are almost ten times the length of the mature caterpillar itself and about half its total weight. *Bombyx mori* uses the silk to spin a cocoon at pupation time; it can spin at a rate of about 6 inches a minute in continuous strands of up to 1/2 mile in length. The silk itself consists of two materials: fibroin, a horny proteinaceous substance that makes up the inner layer, and sericin, a gummy outer covering that is boiled away to give silk its characteristic luster. What appears to be a single thread is in reality a double one; the paired salivary glands each secrete a filament through the minute openings of appendages called spinnerets, beneath the mandibles of the caterpillar. It's spun as a dry fiber, a feat that cannot yet be accomplished by modern technology.

Silk has a number of properties that make it desirable for textile use. Compared to wool and cotton, its density is very low, so it's light and comfortable, not bulky. Its tenacity and elasticity are relatively high and it is therefore highly crease-resistant more so than almost any natural fiber. Silk insulates very well; in its natural state, it probably serves to protect the metamorphosing pupa from environmental unpleasantness. Its insulative properties thus make silk garments warm in winter and cool in summer. It's lustrous and shiny and has high affinity for dyes. It's also highly absorbent, almost three times more so than nylon. Like wool (and unlike synthetic fibers), silk is noncombustible. In proportion to its weight it's the strongest natural fiber known; a silk cable can sustain heavier weights than a metal cable of comparable dimensions. It's elegant, comfortable, and beautiful and has been the most sought-after, the most jealously guarded, and most furiously defended fiber ever.

Silk is such an ancient fiber that its discovery is lost in legend. Ancient literature credits the discovery of silk to Lady Si-Ling, wife of the emperor Huang-ti in the year 2640 B.C. According to one version of the story, Si-Ling was having tea in her garden when the cocoon of a pupating caterpillar dropped into her teacup. In attempting to remove the cocoon from the hot water, Si-Ling removed instead a mass of shimmering fibers. In any event, she was largely responsible for putting the silkworm into production and, in her honor, the fiber was call si, from which eventually comes the English word "silk" and virtually every other name for silk: *sirket* in Mongolian, *seri* in Japanese, *serik* in Arabic, *sericum* in Latin, *seta* in Italian, *soie* in French, *Seide* in German, and *silke* in Swedish.

The secret of silk production, or sericulture, was fanatically protected for centuries. Indeed, silk was the exclusive privilege of the imperial family until 1200 B.C., at which time silk privileges were extended to wealthy nobles and merchants. A class of "silk serfs" arose to keep them well-stocked. When finally, around A.D. 300, the Chinese allowed the release of four Chinese women to Japan to instruct the royal court in the art of sericulture, the Japanese were so grateful that they immediately built a temple in Settsu province in honor of the instructors. In the two hundred years that followed, silk weavers from China moved to Japan in force to found what was until recently the world's largest silk industry.

The western world, meanwhile, was in frustrated ignorance of the secrets of silk. Silk, though not the silkworm (penalty for export of these from China was death), was introduced to the Greeks in the aftermath of Alexander the Great's conquest of Persia and India in the fourth century B.C. With the capture of Babylon, Alexander acquired silken robes of a value of 5,000 talents, roughly equivalent to $7 million. Determined to be more than a ruthless conquerer, Alexander ordered the dispatch of a thousand educated Greeks on his Asian campaign to gather information that would enhance Greek culture. From this thousand came the first account in Western civilization of the silkworm. Although far from accurate, particularly in the details of the silkworm's life cycle, it was considerably better than ideas prevalent at the time ideas that persisted into the Roman era. Even repectable Roman scholars such as Virgil believed that the Chinese "spin their fleecy forests into a slender twine." Dionysius Perigetes, circa A.D. 200, claimed that silk was combed from "the various colored flowers of the land."

The fact that they didn't know what it was didn't stop any Romans from lusting after it. Silk, perhaps because of its exorbitant cost, was the fabric of choice for the Roman aristocracy. At first, its use was restricted by law to the imperial family and meritorious military men but its use soon permeated all levels of society. To put a halt to such business, the Emperor Augustus ruled in A.D. 14 that ordinary citizens were no longer permitted to wear silk, especially "those of lower situations who copied the costumes of their betters." Two years later, an edict was issued forbidding men to wear silk (except as trim) as it was deemed effeminate and decadant. Women of the aristocracy reveled in the luxury of wearing silk, especially the diaphanous gowns from the island of Cos. The passage of the Oppian laws, an attempt to reestablish a moral lifestyle in Rome, restricted the use of silk for ladies' gowns. One account records the reaction:

Not unlike their sisters of every age, the Roman women resented this interference with their freedom and especially in connection with their dress, so they gathered outside the Senate and successfully lobbied for repeal of the obnoxious law, a feminine political action which up to that time was without parallel in history. Thereafter Roman ladies were permitted to dress as they pleased, without restrictions as to silk, color of dye, gold threads, or transparency.

In the reign of Aurelian, A.D. 270 to 275, silk was worth its weight in gold; its price reflected not only its popularity but the difficulty in transporting it from India and the Orient. Most of the silk in Europe was brought from China along the great Silk Road, a trade route extending from China through Turkestan to Iran and eventually terminating at the northeast coast of the Mediterranean Sea. The Silk Road, opened in 126 B.C., was one of the longest roads in the world, stretching over 6,000 miles. Although closed in the ninth century, it was again reopened in the thirteenth century, when Marco Polo used the ancient route to travel to China.

It wasn't until the sixth century that Westerners were able to break the Easterners' secret. In A.D. 555, the emperor Justinian sent two Nestorian monks from the order of St. Basil, longtime residents of China thus privy to the secrets of sericulture, back to China to smuggle out some *Bombyx* eggs. Smuggling was an absolute necessity even for an emperor in view of the death penalty for trading or transporting eggs or caterpillars. The monks managed it by concealing the eggs and mulberry seed inside the bamboo staffs that they as pilgrims were expected to carry around with them. The ruse was successful and as a result all the silkworms in Europe and the United States (until 1865) were descendants of the eggs carried back to Rome by the monks.

Sericulture spread rapidly through Europe. In A.D. 946, King Roger of Sicily conquered Greece and forced a number of silk weavers to settle in Palermo and ply their trade there. Knowledge spread quickly and soon Italy dominated the silk industry in the western world. Weaving and then silkworm raising were introduced to France in the reign of Henry IV (although kings before him attempted to do this Henry II at his 1520 coronation was the first king of France to wear silk stockings). Religious discrimination was in part responsible for the further dissemination of sericulture throughout Europe. In 1685, the revocation of the Edict of Nantes by Louis XIV sent thousands of French Protestants fleeing to Germany, Switzerland, and England. Since most of the silk weavers of France were Protestant, the revocation almost led to the total annihilation of the silk industry there.

England briefly picked up the lead in silk production in Europe. James I did much to encourage its growth he ordered mulberry trees from France and offered free seeds to anyone who would grow them. Outside Buckingham Palace, the king had a mulberry plantation laid out for public display. In 1718, England began manufacturing silk fabrics when John Lombe of Derby, disguised as a worker in an Italian throwing mill, drew pictures of the machinery and returned to England on government subsidy. Lombe subsequently died in great agony, poisoned by an Italian woman sent to England for the purpose.

It was also James I who attempted to introduce sericulture to the New World. In 1609 seven boats set sail for Virginia with mulberry trees and seeds; the eggs and seeds were lost in a terrific storm at sea, a storm of such magnitude it is said to have inspired Shakespeare's play The Tempest. Eventually eggs and seeds arrived, along with a few French people to instruct the colonists, and sericulture was more or less forced on the settlers. The colonial governor ordered that ten mulberry trees be planted on every hundred acres at the risk of a penalty of 20 pounds of tobacco. Ministers extolled its attributes in sermons, politicians enacted huge monetary incentives and poets waxed lyrical:

> *"Where worms and wood doe naturally abound A gallant silken trade must there be found Virginia excells the world at both Envy nor malice can game say this troth."*

Nonetheless, the Virginia planters stuck stubbornly to tobacco, the colonial medium of exchange. The establishment of reeling mills in Pennsylvania stimulated interest in sericulture in the middle Atlantic states and sericulture was conducted on a modest scale throughout Pennsylvania, New Jersey, and Connecticut. At Yale's 1789 commencement, President Ezra Stiles wore a gown made of silk produced from his own silkworms. By the beginning of the nineteenth century, Connecticut was producing between thirty and forty thousand dollars' worth of silk annually.

Silk production took a downturn in the early nineteenth century, at least in part due to a speculative frenzy that gripped wouldbe silk growers. Traditionally, the silkworm, which feeds only on mulberry trees, was raised on white mulberry *Morus alba,* imported from Asia. Neither the species of mulberry native to eastern North America, *M. rubra,* nor the imported and widely grown European species *M. nigra* were especially suitable for raising silkworms. A new type of mulberry, however, was imported into the United States in 1824 by G.S. Perrottet, called *M. multicaulis*. Under the mistaken belief that this tree was used in China for raising silkworms, people became desperate to buy and plant large stands of M. multicaulis. In the heat of what has become known as the multicaulis craze, trees were bought and resold days later for many times their original purchase price. It soon became glaringly apparent, however, that the tree was not hardy enough to withstand northern winters and the bottom fell out of the market; literally millions of trees were simply abandoned by cultivators and would-be silk czars.

In lieu of the production of raw silk, particularly in the aftermath of the multicaulis disaster, many American communities turned to silk manufacture. The government gave this fledgling industry a great assist in 1857 when it increased import duties on silk fabric almost threefold (from 24% to 60%); raw silk, however, could be imported duty-free. The metropolitan Northeast, already firmly established in the textile industry, became the center of silk manufacturing. In Paterson, New Jersey, known as "Silk City," there were at least thirty companies employing 600 people in all aspects of silk manufacture by the mid-nineteenth century. By the beginning of World War I, the numbers had grown to over 24,000 people working in over 300 mills. Paterson was particularly well-suited to the growth and development of the silk industry located near New York, a major distribution center for raw silk, accessible by railroad, and blessed with a water supply peculiarly conducive for dyeing. Many of the mill workers were recent immigrants from Europe who had previously been employed in the silk trades in their countries of origin. Every aspect of silk processing was carried out in Paterson; enterprises engaged not only in production of fabric but also in peripheral businesses such as loom manufacture, dyeing, and even jacquard carding, the creation of punch cards that program looms to weave particular designs. These cards are generally acknowledged to be the conceptual predecessors of computer punch cards. The Jacquard loom was invented in nineteenth century France. Holes were punched in a roll of paper, corresponding to a decorative pattern. This roll pressed up against a series of wire hooks; wherever a wire hook lined up with a hole, it picked up and pulled through a thread. When American Herman Hollerith was faced with the prospect of collecting data for the 1890 census, he adapted the loom design, using electrified wires in place of hooks; when a wire encountered an opening, it transmitted an electrical impulse rather than a thread. This device, a primitive digital computer, was the basis for the formation of IBM.

Labor trouble in Paterson in the early part of the twentieth century led to the breakup of the major mills; manufacturing magnates could not make what they considered a reasonable profit when constrained by union rules and wage scales. Coupled with the invention of synthetic fiber such as nylon in the 1930s, the silk industry faded from prominence not only in Paterson but from the metropolitan Northeast in general. World War II firmly established synthetic fibers in American markets when action in the Pacific theater all but eliminated imports of raw silk from the Far East. Today, however, silk is still very much a valuable commodity, primarily as a luxury fabric (much as it was in its earliest days).

Even without the geopolitical disruptions of World War II, nylon may well have driven silk out of the market; one of the reasons that nylon and other synthetics so quickly replaced silk is that silk is basically a pain in the neck to produce. *Bombyx mori* will eat only the leaves of the mulberry tree preferably *Morus alba,* but it can be coerced into eating a few other *Morus* species. And it eats a lot of mulberry leaves up to 40,000 times its own weight from hatching to pupation. One pound of caterpillars can consume 12 tons of mulberry leaves over the course of a lifetime. *B. mori* begins life as a tiny black caterpillar in Japanese, the first instars are called *kego,* or, hairy babies but within four weeks they grow 25 times larger and 12,000 times heavier. While they're caterpillars, they're susceptible to a number of diseases (one of which, pebrine, proved to have some beneficial consequences it was at least in part due to his investigations of this disease that Louis Pasteur was inspired to formulate his famous theory on the microbial origin of diseases). The mulberry silkworm is effectively a domestic animal no wild specimens of *B. mori* are to be found anywhere in the world and they exhibit the same sorts of physiological frailties that other domesticated animals exhibit.

When the caterpillars are mature, they are transferred to racks where they can spin without interfering with one another. Once the cocoons are collected, they are taken from the farm to a filature, a factory where they are processed. The cocoons are boiled in hot water, which acts both to kill the pupa inside and to remove the outer sericin layer, to give the silk its characteristic gloss. The next step is reeling, or finding the end of the cocoon filament and unraveling and winding it onto a spool. The filaments of several cocoons are reeled together to make threads of particular sizes (e.g., seven cocoons for 20/22 denier silk). Throwing, or twisting threads onto a spool, is followed by soaking and dyeing; silk threads are then reeled into skeins, skeins bundled into books, and books into bales. These bales are then shipped to mills for weaving.

About 350 cocoons are required to make a pair of silk stockings; for one dress, about a pound of silk, or 1,700 cocoons. To obtain 1,700 cocoons, one needs about 125 pounds of mulberry leaves to feed to hungry caterpillars. The 1,700 cocoons require about 10 hours to reel and more time to weave. Sericulture is thus definitely not for the lazy or unenthusiastic.

China leads the world in silk production, its annual output of 30,000 tons of raw silk accounting for 80% of the world's supply. Although for decades Japan led the world in silk production, today it is a distant second to China, producing only about 7,000 tons of raw silk annually. Other world-class producers include India (which produces large amounts of silk but exports very little), South Korea, Indonesia, Brazil, Thailand, and Uzbekistan. Among European nations, Italy takes the lead with an annual output of about 20 tons of raw silk. At the other end of the commercial exchange, the United States is the world's leading importer of silk fabrics and finished goods.

Although the vast majority of silk in the world comes from the cocoons of *Bombyx mori,* other species of insects contribute in a lesser way to the world silk market. Most of the other silk sources are close relatives of *B. mori.* One of the chief nonmulberry silkworms in Asia is *Antheraea yamamai,* the Japanese oak (eri) silkworm. A member of the family Saturniidae, the Japanese oak silkworm spins a silken cocoon comparable in many ways to the silken cocoon of *B. mori.* Unlike the cocoon of the mulberry silkworm, however, the cocoon of the Japanese oak silkworm frequently contains colored pigments (depending upon the light intensity experienced by the larva); pigment differences, as well as some structural differences in the silk itself, mean that dyeing conditions are more critical. Moreover, with current rearing practices, only one generation of the Japanese oak silkworm per year can be produced. Other saturniids have been examined periodically as potential commercial silk sources; in Turkey, the moth *Saturnia pyri* has been proposed as a silk source preferable to the mulberry silkworm because of its polyphagous habits and its larger cocoon.

The search for a better method of silk production is not without its own pitfalls, however. In the mid-nineteenth century, an amateur lepidopterist named Leopold Trouvelot living in Cambridge, Massachusetts, imported eggs of several moth species in an attempt to develop by hybridization a silk-producing moth resistant to the many diseases that afflict the mulberry silkworm in culture. Through an oversight, caterpillars of one species, *Lymantria dispar,* accidentally became established in his neighborhood. As caterpillars, *L. dispar* disperse by ballooning, or spinning a thin thread of silk, which provides the tiny first instars with lift and buoyancy. Thus, these so-called gypsy moth caterpillars can cover a great deal of territory. From Trouvelot's backyard, the gypsy moth went on to colonize all of New England by the end of the century, reaching epidemic proportions in some forested areas. In less than a hundred years, it had expanded its range from coast to coast, aided in its American conquest by the tendency of the flightless females to lay masses of several hundred eggs in the wheel wells of campers and other vehicles, which transported the species across the country. As a major defoliator of dozens of forest tree species, ranging from white pine to black oak, the gyspy moth has made an economic impact far different from that envisioned by the well-intentioned but misguided Trouvelot.

Another group of arthropods of some importance in commercial silk production are spiders. Unlike caterpillars, spiders spin silk not from a pair of salivary glands but rather from a series of silk glands of various descriptions in their abdomen. Silk is pulled out through openings in the spinnerets, which vary in number depending on the family, either by the legs or by the entire body, after the spider anchors down one end to the ground. Virtually all spiders spin silk, which plays a vital role in many aspects of spider biology. Different types of silk are spun for different purposes. For example, even within an orb web, sticky strands serve to snag prey while smooth strands are used by the spider to negotiate its own web and so avoid entanglement.

Depending on species, spider silk varies in thickness from 1/50,000 inch to about 1/5,000 inch; in comparison, a blonde human hair, at 1/250 inch, is a veritable tree trunk. Elastic and resilient, some spider silks can be stretched over 20% of their full length. Historically, spider silk was used in constructing fishnets, bird snares, or woven into small pouches; cobwebs in particular have found use as bandage material (to stanch blood flow) and even, stretched and doubled over, as artists' canvas. The most commercially important use of spider silk today is as material for constructing crosshairs in optical instruments, including surveying equipment, range finders, microscopes, bombsights, and telescopic gunsights. For gunsight crosshairs, silk from the immature stages of *Latrodectus mactans,* the black widow spider, is desirable; despite the potentially dangerous effects of handling this spider, which as an adult is capable of inflicting a fatal bite, some brave souls harvest silk from semidomesticated specimens.

At least partly because of the difficulties attendant upon collecting sufficient quantities of silk from potentially lethal spiders, the U.S. Army has funded research aimed at producing silk in the laboratory by genetic means. Genes that encode different silk proteins have been cloned into bacteria, which then manufacture these proteins in mass quantities. Future military applications of silk fibers, once they can be manufactured cheaply and rapidly, include incorporation into bulletproof vests, lightweight helmets, and parachute cords. Spider silk is particularly well-suited for this last application due to its extraordinarily low "glass transition temperature," or temperature at which the fiber becomes brittle. Because it is still elastic and resilient at temperatures as low as -40°C, silk fiber strands have a lower risk of breaking in freezing weather than do most other fibers.

Stored-product Pests and Home Invaders

Along with all the good things that resulted from the domestication of plants for food came a few brand-new problems. By storing food, people were unintentionally revolutionizing the life of insects that had heretofore made their living feeding on foodstuffs that were typically few and far between. Many arthropods inhabited the nests of birds and mammals, particularly rodents, and fed on the grain caches assembled by their unwitting hosts. Others were associated with dead animals and dried animal carcasses, never a resource in tremendous supply, and still others lived on whatever they could find in leaf litter or soil. For these arthropods, the human practice of stockpiling food created a bonanza beyond their wildest imagination. From the human point of view, stockpiling food and other materials created a monster in the form of storedproduct pests. For virtually every kind of product stored in any quantity by humans, even manufactured goods, there are insects that eat it, almost to the exclusion of anything else.

Probably the most numerous and economically destructive stored-product species are those that feed on grain and grain products. That this mooching has been going on for some time is evidenced by the fact that remains of stored-product species, such as the beetle *Tribolium castaneum,* have been found in alabaster vases in Tutankhamen's tomb dating back to around 1350 B.C.*Sitophilus granarius,* another grain pest, has been found in grain dating back to the ninth to seventh centuries B.C. in Israel. In fact, one minor grain pest, a colydiid beetle called *Thaumasphrastus karanisensis,* was first discovered in grain from an Egyptian tomb; it was subsequently found breeding in south Texas in a rice mill. Many pests of stored grain actually do feed on grain in the field in places where the weather is nice enough to allow it. There are dozens of species, however, that make their living in mills, elevators, and kitchen pantries today.

Stored-product insects differ in their approach to raiding larders. Some, like the weevils, prefer to feed on whole grains; these so-called primary pests include the grain weevils, the grain borers, and the bruchids, as well as some species of Lepidoptera (like the Angoumois grain moth). These are mostly pests in granaries, elevators, corn cribs, barns, and in any other place where intact kernels are stored. Others prefer to feed on processed or damaged grains; these secondary pests include the flour beetles and mealworms. These grubs are often called "bran bugs" and are pests in flour mills and in processed food. Many species of Lepidoptera are also secondary pests, particularly those in the Pyralidae, or snout moth family. These include the Indian meal moth *Plodia interpunctella* and the Mediterranean flour moth *Anagasta kuehniella.*

One characteristic that most grain-feeding stored-product pests share is an extraordinary tolerance to dry, hot conditions. The Mediterranean flour moth *Anagasta kuhniella* can develop on foods with less than 1% water content without ever taking a drink of water. Many stored-product species meet their dietary requirement for moisture simply by extracting water from the carbohydrates in their food. Mealworms, many of whose relatives in the family Tenebrionidae live in the desert, can actually absorb moisture from the air, even when the relative humidity is below 50%, to supplement their water intake. Adapting to feeding on a constant and predictable food supply has led in some species to the loss of a means of dispersal, or getting around; there's no need to. The granary weevil, *Sitophilus granarius,* for example, has only rudimentary vestiges of what once were functional wings.

In some ways, when humans sieve, winnow, chop, blend, and otherwise process plant material, they render it far more to the liking of insects, who don't then have to contend with a lot of indigestible chaff. Processing can also remove a lot of the nasty poisons that plants tend to accumulate to discourage herbivory. But not always, and many stored-product species have a remarkable tolerance for what would otherwise seem to be totally unsuitable food. The cigarette beetle *Lasioderma serricorne* (Anobiidae) has an extraordinary predilection for cigars, cigarettes, and chewing tobacco, even though the tobacco plant, in the field as well as chopped and rolled, is loaded with nicotine, a substance that is lethal to most insects. Drugstore beetles (*Stegobium paniceum*) have been found happily breeding in jars of cayenne pepper, a concentrated diet of which is certainly beyond the capabilities of even the most diehard pepper fans. And confused flour beetles (*Tribolium confusum*) have been found contentedly infesting bales of marijuana confiscated and stored by the Federal Drug Enforcement Administration (although the name "confused" refers to the fact that the species is often confused with *T. castaneum,* the red flour beetle, and doesn't refer to any diet-related mental state in this instance).

9

Eating Insects in Life Science

Introduction

Predators other than insects also make a diet out of insect prey; insects represent an abundant source of high-quality protein that is relatively easily harvested relative to larger, stronger forms of animal protein on the hoof. Among the most fervent predators of insects are their relatives the arachnids, but insects provide an important source of nutrition to all classes of vertebrates, including fish, amphibians, reptiles, birds, and mammals.

Entomophagous (or insect-eating) predators in general, be they arthropod or nonarthropod, share certain anatomical features that facilitate the capture and consumption of insects. For one thing, they generally tend to be larger than their prey. (One exception to this pattern are the ants, which, by foraging collectively in organized groups, can manage to tackle prey much larger than they are individually. Army ants, for example, forage from bivouacs containing up to 20 million individuals weighing a collective 20 kilograms or more). In addition, predators of insects generally possess welldeveloped sense organs for detecting the presence of their prey. Visually orienting dragonflies, for example, possess enormous compound eyes with over 20,000 facets, which provide them with virtually 360° vision. Most predators are well-equipped for traveling long distances, since insect prey are rarely abundant enough in any one locality to support a large predator population; they tend to be excellent fliers or runners. One conspicuous exception to this pattern are predators that concentrate on consuming sedentary prey such as aphids, which, with their asexual reproduction, can build up locally to enormous population sizes. Finally, most predators of insects are equipped with appendages to assist them in seizing and securing their prey. Modified mouthparts as well as legs play a role in this aspect of predation. Some species greatly reduce the need for physically grappling with prey by injecting paralytic venoms, thus allowing them to tackle prey considerably larger, stronger, and heavier than they are themselves.

Predators (as opposed to other types of carnivores) typically kill their prey, usually consume it themselves (as opposed to killing it and giving it to their offspring), and kill more than one prey individual over the course of their lifetimes. Arthropod predators employ a variety of prey-securing strategies that involve different amounts of energy input. Some predators eliminate the search and pursuit aspect of predation by simply concealing themselves and waiting for prey to come to them. In other words, they set a trap and wait. Spiders are typical sit-and-wait predators. Orb weavers (in the family Araneidae) construct large circular webs that act as aerial filters that intercept and snare prey. The sheet webs of agelenid spiders operate on the principle of flypaper and capture almost exclusively insects that crawl and walk carelessly. Trapdoor spiders construct an underground trap with a silken hatch; when a passing insect wanders by and stumbles in, the trap door is pulled shut by a silken thread and the hapless insect is confined with no avenue for escape.

Praying mantids are also ambush predators, which move in the general direction of a potential prey and wait, poised and ready to strike, until it blunders into the vicinity. Mantids (like many ambush predators) are morphologically well-equipped for a sudden strike. They have raptorial, or grasping, forelegs. These legs are so specialized for snagging prey that they are essentially useless for walking. The segment of the leg nearest to the body (the coxa) is greatly elongated and the next segment is serrated and grooved along its lower surface. The spines of the next segment in line fit snugly into the groove when the leg snaps shut, presumably around the body of a hapless meal. The elongation of the coxa causes the mantis to hold its legs folded underneath its body, as if in prayer hence the name "praying mantis" (and not, as it's often incorrectly but not inappropriately written, "preying mantis"). The praying mantis is unusual among insects in the degree to which it can swivel its head around; if insects can be said to have shoulders, mantids are unique in their ability to look over them.

Both of these traits the beatific attitude and the disconcerting stare have attracted considerable attention to mantids worldwide. The sudden and forceful strike of the praying mantis which takes only fifty to seventy milliseconds to complete has actually inspired an entire style of kung fu martial arts in which many of the postures and movements mimic the hunting behavior of the praying mantis. The style was founded by a man named Wang Lang almost 400 years ago in the province of Shantung. Inspired while observing a praying mantis dispatch a cicada, Wang Lang, who was already a renowned martial artist, refined a series of movements, hand and arm positions, strikes, and grabs based on subsequent studies of mantis predation. The success of his style, demonstrated in several bouts with the world-famous Shaolin monks (who were devoted students of martial arts), led to the widespread adoption and subsequent modification of the praying mantis style.

One of the drawbacks of a sit-and-wait approach is the fact that there may be long periods of waiting with no success. Many sit and wait predators, accordingly, have a tremendous tolerance for starving. The spider *Linyphia triangularis* can develop, depending on the amount of food it receives, from egg to adult in anywhere from 70 to 200 days and can weigh anywhere between 2 and 25 milligrams when full grown, depending on the availability of food.

Some sit-and-wait predators enhance their chances of success by not only setting a trap but baiting it as well. *Mastophora* spiders are orb weavers that do not spin an orb web. Rather, they spin only a single thin silken strand equipped with a sticky ball suspended at the end. The spider dangles the ball with one front leg and tosses it at passing insects when they approach. This ball is imbued with spider-synthesized analogues of the female sex pheromones of neighborhood moths; thus, these spiders feed almost exclusively on male moths. The habit of these spiders of tossing their sticky ball at passing insects has earned them the common name of bolas spiders, in honor of the lariat-like bolas used by Argentine gauchos to lasso their prey. In a more scientific vein, the pitching skills of one species earned it the specific epithet *M. dizzydeani,* in honor of the legendary baseball player Dizzy Dean.

Insinuation is the term given to the ability of certain predators to lull their prey into a false sense of security. Insinuation is behind the bait-and-capture technique of *Salyavata variegata* (Reduviidae), the termite-eating assassin bug. The bug captures a termite, sucks it dry, and approaches the colony while carrying the carcass. When a worker approaches to recycle the body, the reduviid backs off until the worker is far enough away from the colony that the reduviid can attack without risk of provoking soldier termites to rescue their nestmate.

Cleptoparasitism is a mode of predation in which search and pursuit are carried out by one species and consumption by another in other words, some predators simply steal prey from other predators. Many species of spiders are cleptoparasites that set up housekeeping within the webs of larger spiders and steal prey before the resident can secure it. *Nephila clavipes* builds a large orb web with barrier webbing in front of the capture surface. Smaller spiders live in the barrier webbing and steal unwrapped prey. Although there may be as many as three or four cleptoparasites living in a web, the resident *N. clavipes* seems unaware of their presence. *Pholcus phalangioides,* the cellar spider, while a cleptoparasite that frequently raids the nests of other species to steal prey, also lures resident spiders out of their lairs by plucking web strands in such a way as to suggest that it is itself a struggling victim; when the resident spider rushes out to snare the purported prey, the cellar spider pounces and consumes the would-be predator.

Only a little more energy-intensive than sitting and waiting for prey is searching for and attacking relatively sedentary prey. Softbodied homopterans such as scale insects or aphids lead a remarkably sedentary existence; moreover, they are small and relatively defenseless. Accordingly, a number of predaceous insects are specialized for feeding exclusively on them. Such predators include ladybugs, the larvae of hover flies, the brown and green lacewings, and even a few caterpillars. Unlike predators of active prey, predators of soft-bodied slow-moving prey generally do not use specific signals visual, acoustical, or chemical for locating their prey. Rather, they rely on random searching movements to bring them into the vicinity of a large population, on which they may feed at their leisure.

A tremendous variety of insect predators actively search out and pursue their prey. In the air, dragonflies are a prime example, as are robber flies (true flies in the family Asilidae); on the ground, ground beetles and tiger beetles are typical active predators. Many of the ground-dwelling arachnids obtain prey in this fashion; these include the wolf spiders, scorpions, the sun-scorpions, and the wind-scorpions. In aquatic habitats, diving beetles, whirligig beetles, and belostomatids (giant water bugs) are fast swimmers that chase down their prey. Once large active prey are captured, predators must subdue and secure them. Although physical strength in many cases can suffice (as in the case of mantids), many predators of active prey use paralytic venoms to immobilize their prey and reduce the energy expenditure involved in securing and subduing them. While they are themselves large as wasps go, cicada killers (wasps in the family Sphecidae) would have considerable difficulties transporting their prey, which weigh as much as or more than they do, back to their underground burrows if they were kicking and struggling the whole way; these wasps inject a paralytic venom that instantly immobilizes the cicada not only for the duration of the trip back to the burrow but also throughout the entire developmental period of the grub, which feeds on the still-living prey.

Parasitoids

"Parasitoid" is a term coined over eighty years ago to describe insects that are parasitic only during their immature stages; adults are free-living. Those attacking invertebrates almost always kill their hosts and thus should more properly be considered predators, since parasites don't usually kill their prey whereas predators invariably do. Hence the term "parasitoid" a word to describe an insect lifestyle that has no close analogy in any other phylum. Unlike predators, adults of these species locate prey for their offspring, not for themselves; they like as not content themselves with sipping nectar from flowers.

The vast majority of parasitoids are found in only two orders: Diptera and Hymenoptera. Parasitoids vary in their degree of intimacy with their hosts. Some parasitoids lay their eggs on the body of the host, where the eggs attach and develop externally. Others lay their eggs on leaves; the eggs hatch when they are swallowed by the host and the larvae develop inside the body of the host, consuming bits and pieces of internal organs at their leisure. Still others, the endoparasitoids, lay their eggs directly into the body of the host, so their offspring never have to venture outside the safety of a nice, cozy insect body cavity until they are ready to pupate at which point they usually burst through the body wall and kill their still-living but much depleted host.

Parasitoids share certain ecological similarities with predators. While as larvae they are sedentary, as adults they must search for widely dispersed hosts in which to lay eggs; thus, parasitoid females are well-equipped to travel considerable distances and to locate prey by acoustical, visual, or chemical means. A parasitoid of a gall fly maggot, for example, forages along the stem of the fly's host plant and waits for vibrations resulting from the feeding movements of the maggot mouthhooks. Parasitoids of the cotton bollworm orient to aromatic chemicals released when the bollworms chew on their host plants. Wasp species that parasitize leaf-mining caterpillars can detect leaf mines visually and orient to them specifically in order to locate hosts. Also like predators, parasitoids frequently use paralytic venoms to immobilize their hosts (as in the case of species that deposit eggs on the outside of their hosts). In the case of endoparasitoids, other substances (possibly viruses) are injected that arrest the development of the host so that the parasitoid can complete its development before its host undergoes metamorphosis.

On the other hand, parasitoids also share ecological characteristics with parasites. While predators frequently are highly polyphagous, parasitoids, like many parasites, tend to be highly host-specific. One possible reason for such specificity is that, like parasites, parasitoids must synchronize their development relatively precisely with that of their hosts. They also tend to be smaller than their hosts, whereas predators are often considerably larger than their prey. At least partly due to their small size, parasitoids also tend to have far greater reproductive capacity than do predators, a characteristic shared with parasites.

Biological Control of Insects

Biological controls is the use of the natural enemies of an economic pest species for the purpose of reducing the population size of that pest ideally to the point where the pest no longer has a negative economic impact. In most cases, the pest species is a pest because it's an exotic (introduced from another locality) and in its new habitat, far from its normal contingent of enemies, its populations can grow large and destructive rather quickly. The natural enemy is usually obtained from the pest's native habitat. Traditionally, insect "natural enemies" include parasitoids, predators, and pathogens. Use of these natural control agents falls under the rubric of "classical," or conventional, biological control.

Classical biological control is by no means a new concept. In the thirteenth century and earlier, nests of *Oecophylla smaragdina* (weaver ants) were placed in citrus and litchi trees by farmers in China to control plant-feeding stink bugs (*Tessarotoma papillosa*) and other pests; nests of pharaoh ants (*Monomorium pharaonis*) were brought into barns in China during the same time period for controlling stored products insects. The first long-distance transfer of an exotic predator took place in 1762, when mynah birds from India were brought to Mauritius to control grasshoppers. In more contemporary times, the pest-control crisis that launched biological control into scientific circles was an infestation in California of cottony-cushiony scale on citrus. *Icerya purchasi* was first noticed in Menlo Park around 1868; by 1885, it was a threat to the new citrus industry. C.V. Riley, then State entomologist for Missouri, suggested in 1872 that the pest was Australian in origin; by 1878, as chief entomologist for the USDA, he sent Albert Koebele to Australia to look for natural enemies in the home territory of the cottony-cushion scale. Koebele was sent officially, however, as State Department representative to the International Exposition in Melbourne. Since no foreign travel was allowed for USDA employees Koebele's visit was tacked on as a rider to an appropriations bill just to accommodate Riley.

In Australia, Koebele found a predaceous ladybug, *Rodolia cardinalis,* known as the vedalia beetle. Stocks were sent to Los Angeles to D.W. Coquillet, who reared them and released them onto scaleinfested trees enclosed in tents on J.W. Wolfskill's Los Angeles farm. From there they spread to orchard trees nearby. People came from all over the state with little boxes to collect the beetles and take them home. With the assistance of citrus growers and the Los Angeles County Board of Horticultural Commissioners, the beetles were soon established throughout the county. Within one year of their introduction, shipments of citrus went from 700 to 2,000 box cars, the citrus industry was saved, and, best of all, $500 remained of the $2,000 originally allocated for Koebele's trip to Australia. Koebele received a gold watch and his wife a pair of diamond earrings from the grateful growers.

The spectacular success of the vedalia venture catapulted the approach, clubbed "biological control" by H.S. Smith in 1919, into prominence. Examples of successful biological control programs are now legion. The majority involve accidentally introduced pests that form large populations. Many of these targets for biocontrol are relatively sedentary pests (e.g., aphids and scale insects). Most of the successful control agents are parasitoids, at least in part because more parasitoids have been tried in biocontrol programs; there are currently over 1,700 intentionally introduced and established parasitoids, compared with fewer than 500 predators, in the U.S. Host specificity may be the key to a successful biocontrol effort; many of the most successful predators, for example, are exceptionally specialized in terms of range of feeding preference. Ladybugs, for example, feed almost exclusively on aphids or scale insects and include some of the more spectacular biocontrol success stories. Biocontrol agents are now widely available to home gardeners. Both predators and parasitoids can be obtained through mail-order catalogs and arrive complete with instructions for rearing and release.

Insects as Prey (Studies in Bad Taste)

The Main Problem faced by any would-be consumer of insects is the interest displayed by virtually all species in self-preservation (i.e., not being eaten). Almost every species of insect has some sort of defense against predators. There are some apparent exceptions: periodical cicadas, for example, which emerge en masse every thirteen or seventeen years seem to lack any kind of morphological or behavioral defense against predators. They may be able to get away with this apparent lack of means for protecting themselves because at least a few individuals escape notice in the mass confusion of a mass emergence; predators quickly eat their fill of cicadas and leave the remaining individuals alone, defenseless though they may be. This strategy is often referred to as predator satiation.

There are several basic defense strategies among insects. Some defenses are in operation all the time, irrespective of whether a predator is nearby. Usually, such defenses are built in as part of the body plan and generally act to reduce the probability that an insect will be discovered by an enemy. Built-in concealment can take several forms. Crypsis, or camouflage, is quite common. In fact, insects are quite remarkable in this regard. There are insects that rest on tree bark that are indistinguishable from tree bark, and insects that rest on rocks that look just like rocks. Crypsis can even be a dynamic process. *Misumena* crab spiders can change color, from yellow to white, to match the color of the flowers on which they sit. There's even a species of grasshoppper that lives in the desert and is basically sand-colored but when a fire burns through the area, the grasshopper develops black "scorch marks" on its body to enhance its resemblance to the fire-strafed environment.

Background matching can, however, be a tricky business if the environment is subject to change and there is no corresponding flexibility in appearance. The peppered moth is a case in point. Butterfly collectors, at times regarded as eccentric, are if nothing else attentive to variation; they go to great lengths to find rare and unusual forms of species. In the nineteenth century, it was considered a great prize in England to collect a peculiar form of the peppered moth, *Biston betularia carbonaria,* a darker version than the usual form *B. betularia insularia.* But after 1845 or so, the dark forms kept getting more and more common and, within 100 years, in Birmingham, England, 90% of all the forms collected were dark. H.B. Kettlewell decided to investigate the situation. The light form of the peppered moth rests during the day on lichen-covered bark; since it is cryptic and resembles lichen-covered bark, it is almost invisible and is rarely eaten by birds. But when Kettlewell released both dark and light moths around Birmingham, the only ones that weren't eaten were the dark ones. Birmingham is in the middle of the industrial region of England. Soot, smoke, and general pollution covered the trees, not only darkening the trunks but also killing the lichen. Light moths on a sooty trunk were literally dead giveaways and got picked off by birds, while the dark ones, formerly rare because they were conspicuous on light-colored trunks, survived far better in industrialized areas. In Dorset, a rural area with no pollution to speak of, the light forms were more common and, when Kettlewell repeated his experiment there, survived better in the presence of birds and other predators than did the dark ones. Pollution control measures implemented in the 1970s were accompanied throughout the Birmingham area by a resurgence in the frequency of the light-colored peppered moths.

Aside from crypsis, there is another form of concealment known as homotypism, or resembling something inedible. There are insects that resemble bird droppings, which presumably no selfrespecting bird would deign to sample; there are even chrysomelid beetles that resemble caterpillar droppings (if possible, an even less appetizing prospect). Three groups of insects are particularly renowned for their reliance on resembling anything but an insect. Many of the treehoppers (Membracidae) look for all the world like thorns and tend to rest on the stem in places where thorns are likely to be found. Most phasmids, or walking sticks, as the name suggests, bear a strong resemblance to sticks or twigs. Many species of caterpillars in the family Geometridae, called inchworms, also resemble sticks or twigs. Both phasmids and inchworms have behavior consistent with their appearance they can remain motionless, even if prodded, for hours on end. Any twig mimic that begins crawling away after it's touched is likely to lose its credibility (as well as its life). Homotypic insects, like cryptic insects, can have some measure of flexibility in terms of changing their appearance to conform with changes in the environment. In Arizona, an inchworm caterpillar (*Nemoria arizonaria*) feeds on oak catkins (flowers) in the spring and on oak foliage in the summer. In the spring the caterpillars look like catkins and in the summer they look like oak twigs. The change in appearance is apparently due to a physiological response to chemicals in the host plants. Catkins are low in chemicals called tannins; leaves and twigs, in contrast, have high levels of tannin. Caterpillars that consume a low-tannin diet take on the catkin form, while caterpillars that consume a high-tannin diet take on the twig form.

Sometimes the means for looking inedible are obtained from the environment. Some species of green lacewings place bits of plant material or other debris on their backs, so they resemble little traveling dust mice. Larvae of *Chrysopa slossonae,* a green lacewing, prey on woolly alder aphids, which are protected by pugnacious ant bodyguards. These ants consume the honeydew produced by the aphids and zealously protect their food source from marauding would-be predators. To avoid detection by the ants, the larval lacewings pluck the waxy cottony fluff from the backs of woolly alder aphids and affix it to specialized "grappling hooks" on their own backs. Thus disguised, they can wander in among the ants without detection a real-life example of a wolf in sheep's clothing. Another example of the use of environmental debris for protection hits closer to home (at least, to some homes). The masked bedbug hunter (*Reduvius personatus*) is an assassin bug that eats bed bugs. The immature nymphs are covered with a sticky substance that attracts house dust, which soon covers them from head to tarsus (making, no doubt, for some unnerving confrontations for homeowners unaccustomed to dust that travels on its own power).

One other primary defense is not only to look inedible but actually to be inedible. Bad taste in insects can arise in at least two ways. Insects can either manufacture poisons themselves or they can store up (sequester) poisons from the plants they eat. Almost invariably associated with bad taste is aposematism, or warning coloration. Insects that are distasteful, instead of attempting to conceal themselves, actually advertise the fact with unmistakably bright colors. Curiously, the colors assumed by most aposematic insects are the very same colors used by humans to warn of potential danger red (like stop signs), yellow or orange, and black (like highway warning signs). Thus, most venomous bees, ants, and wasps (Hymenoptera) are brightly colored, often yellow-and-black striped.

Not all brightly colored insects, however, are distasteful. Some are cheats, that is, perfectly edible species cashing in on the fact that predators are loathe to attack brightly colored prey, particularly after a bad experience or two. Such an arrangement is called mimicry. Batesian mimicry, named for H.W. Bates, a nineteenth-century naturalist, is the situation in which a palatable mimic resembles a toxic model. Harmless perfectly edible insects in many orders mimic the venomous bees, ants, and wasps everything from flies to beetles to even moths. Hover flies in the family Syrphidae are particularly accomplished in this regard. Not only do they have the coloration pattern down perfectly, they even make their stubby fly antennae seem more wasplike by dangling their thin black forelegs in front of their face, where long, black antennae would be expected to occur. The master mimics tend to appear either early in the spring or in late summer, avoiding the period when naive young fledgling birds who don't know enough to avoid yellow-and-black striped insects are first encountering stinging hymenopterans and learning their lessons.

Many defenses are activated only in the presence of a predator. Retreat and hiding are simple but effective responses to the presence of a predator (it's the approach used by cockroaches when you turn your lights on in your kitchen). There's also a distinctive behavior called thanatosis playing dead. Many predators can detect and capture only living prey, so by playing dead a potential morsel can make a predator lose interest and look elsewhere. This defense is popularly known as "playing possum" but, considering that most of the 20,000-plus members of the family Chrysomelidae (leaf beetles) resort to this strategy when disturbed, it probably should be called "playing beetle." In this deathlike state, many beetles can remain motionless, even when handled, for several hours.

Many species depend on the element of surprise to disconcert or frighten off a predator. These species tend to be cryptic until disturbed, whereupon they flash a startling pattern in the face of their enemy. One very common flash pattern is an eyespot, a marking that resembles an eye (even down to the glint). A bird, thinking it's about to snag a meal, suddenly finds itself looking into what seems to be the baleful stare of a very large animal. Allen M. Young, a tropical biologist, has an interesting theory to account for the distribution of eyespots in tropical butterflies. He noticed that eyespots are very common in butterflies that feed on rotting fruit on the forest floor but relatively rare among butterflies that feed on nectar. He suggested that the butterflies eating fermenting fruit actually become intoxicated and impaired to the point that, if they became aware of a predator, they would need a few seconds to sober up and coordinate themselves to fly away. The eyespots, flashed in the face of the predator, give the bibulous butterflies the few seconds they need. More abstemious species ostensibly have their wits about them at all times and don't need the extra time to gather their wits about them, as it were.

Even in the insect world the best defense is often a good offense. Many species retaliate when attacked. Since insects are by and large not very large, brute strength is rarely an effective means of fending off an enemy. Most insects resort to chemical weapons for defending themselves against attack. Defensive compounds are usually manufactured and stored in cuticle-lined glands that are equipped with a delivery system allowing the insect to discharge the contents with maximum efficiency. Defensive glands come in an almost endless variety of types. Oozing glands discharge their contents slowly as liquid droplets, which may gradually flow over the body's surface. The aptly named stink bugs (Pentatomidae) produce a mixture of over a dozen different vile-smelling substances from paired oozing glands on their thorax. Eversible glands are concealed inside the body until the insect is disturbed; upon disturbance the gland is everted, or popped out, by hydrostatic pressure. Caterpillars in the family Papilionidae all possess a Y-shaped eversible gland that tucks in behind the head when not in use. When a caterpillar is poked or grabbed, the gland (called an osmeterium) pops out and discharges its odoriferous contents (which in some species include such compounds as isobutyric acid, the chief odor principle of rancid butter).

Perhaps most spectacular of all are reactor glands, actually functional chemical laboratories in which two or more relatively inactive compounds are combined to create a chemical reaction. The bombardier beetle owes its name to its ability, when disturbed, to eject a hot liquid from its abdomen several times in succession with such force that an audible popping sound is produced at each discharge. As described by naturalist John Westwood in 1839, "These beetles on being seized immediately play off their artillery . . . burning the flesh to such a degree that only a few specimens can be captured with the naked hand." The reactor gland of the bombardier beetle is two-chambered; the contents of one chamber are moved into the second chamber, where they combine with other materials in a heat- and oxygen-generating explosion. The heat moves the temperature of the discharged liquids to the boiling point and the oxygen gas liberated propels them outward. The hot liquid is released in a series of rapid spurts, averaging 500 pulses per second, rather than in a constant stream. Rapid muscular contraction controls the flow of materials from reservoir to chamber. Overall, the whole system is eerily similar in design to the V-1 buzz bombs used by the Nazis in London during World War II. The buzz bomb, too, was constructed in a multiple-chamber design. It owed its name to the pulsing sound it emitted when set off; the pulsing in turn resulted from the action of an oscillating flapper valve (analogous to the contracting sphincter muscles) that slammed shut after kerosene from one chamber entered into a second chamber and ignited. The force of the explosion pushed the exhaust gases out of the chamber, propelling the bomb forward and starting the cycle over again.

In some species of insects, toxic substances are not stored in glands but instead circulate freely in the blood or hemolymph. These insects, when attacked, bleed to protect themselves in a process known as reflex bleeding. There are specialized weakened areas of cuticle through which an insect can force its blood, often with sufficient force to send it spraying over several inches. Many beetles are capable of reflex bleeding, ladybugs and oil beetles among them. Reflex bleeders are capable of withstanding losses of up to 25% of their total blood volume without suffering adverse effects.

Physiological defenses are the last resort of an insect after the enemy has penetrated its body. These are generally initiated in response to attack by parasitoids. Hemocytes, or blood cells, of the host congregate around the developing parasitoid larva; more adhere and eventually a capsule consolidates. In one sort of capsule, a black pigment called melanin is deposited, cutting off oxygen to the parasitoid. There are countermeasures taken by parasitoids, however. Some parasitoid fly maggots develop a respiratory tunnel that connects to the host tracheae so that they have a constant supply of oxygen, despite being sealed into an airtight capsule.

Insects as Medicine (Flies in the Ointment)

Despite the impressive array of toxins insects produce (or perhaps even because of them), humans have been swallowing insects -deliberately for centuries. Ever optimistic and opportunistic -people have throughout the ages reasoned that there must be some way to turn insects to their advantage. Perhaps the most articulate expression of this belief that insects exist for the material benefit of humans can be found in a book titled *Insectotheology, or A*

Demonstration of the Being and Perfection of God from a Consideration of the Structure and Economy of Insects, published in 1699:

> *Every event in nature proclaims that whatever happens is directed by a Being infinitely wise. From this principle it follows that God hath determined the uses of animals in general and of insects in particular, and that he hath destined them to serve, some one purpose, some another. It is not chance, then, that hath made insects useful, but the eternal decrees of providence; and man hath only turned to his use what was originally intended for his service.*

No other virtue coming immediately to mind to those charged with divining these purposes, the most reasonable explanation for the existence of insects more often than not had to do with maintaining human health.

Experimentation with insects as curative agents is one of the more ancient scientific pursuits. The Ebers Papyrus, an Egyptian medical treatise that dates back to the sixteenth century B.C., contains references to insect remedies considered ancient even in those days. Etymologically speaking, the very word "medicine" has entomological roots it derives from the same root as does the word "mead," or fermented honey wine, to which were ascribed remarkable curative properties.

Divine powers are notorious for working in mysterious ways and it has been the lot of humankind over the centuries to ascertain precisely which medicinal powers were bestowed on which insect. The most optimistic presumed that the Supreme Being in his infinite wisdom not only created all things for human use but invested in all things some physical sign, such as shape or color, to demonstrate its particular usefulness. Thus was formulated what has come to be known as the doctrine of signatures, formalized in Latin as *Similia similibus curantur* loosely, "Let likes be cured by likes." Botanical examples of the application of the doctrine of signatures include the use of trilobed liver-shaped leaves of *Hepatica* against jaundice, a liver ailment. Entomological examples, however, are by no means difficult to discover. Conspicuously hairy insects, for example, quite often found their way into hair tonics and baldness cures. In his *Historia Naturalis*, Pliny the Elder quoted Varro as saying:

The heads of flies, applied fresh to the bald place, is a convenient medicine for the said infirmity and defect. Some use in this case the bloud of flies; others mingle their ashes with the ashes of pear used in old time or els of nuts; with this proportion, that there be a third part only of the ashes of flies to the rest, and herewith for ten days together rubb the bare places where the hair is gone. Some there be, again, who temper and incorporate the said ashes of flies with the juice of colewort and brest-milke; others take nothing thereto but honey.

Bees, too, are noted for hirsuteness and, not surprisingly, appear in various preparations designed to alleviate the embarrassments of hairlessness. Dr. James, in his *Medicinal Dictionary* (1743) writes, "If this powder [bee salts] is mixed in unguents with which the head is anointed, it is said to cure the Alopecia and to contribute to the growth of hair on bald places."

In still another application of the doctrine of signatures, singing insects crickets, katydids, and some grasshoppers were frequently recommended for disorders of the throat and ears. Throughout Europe during the Middle Ages, the body juice of crickets was regarded as a sovereign remedy for ear and throat difficulties. The practice may even date back to biblical times, when Jewish women were said to carry *chargol* locusts (the precise taxonomic identity of which is hotly contested among scholars) in their ears to prevent earache. Another singing insect, the cicada, found widespread use in China and Japan as a constituent of earache cures and, curiously, was used in India among the Santals in ointments mixed with male crickets (which, in most species, are the only ones that can sing) to cure screaming fits.

Earwigs, too, were prescribed for those with hearing difficulties, perhaps due to the recognition of the fan- or ear-shaped hind wing (whence the name "earwig") as a signature. From Dr. James:

Oil of Ear-wigs is good to strengthen the nerve. These insects, being dried, pulverised, and mixed with the urine of a hare, are esteemed to be good for deafness, being introduced to the ear.

Often, signatures or signs are derived not from appearance but rather from behavior. For example, the widespread use of all manner of insects to enhance fertility or virility may have resulted simply from observation of the tremendous powers of increase of insects. Termites in particular, which can form colonies with populations upwards of 1 million individuals, have caught the attention of individuals seeking to restore lost virility or strength; the debilitated

Surjee Rao, prime minister of Scindia, chief of the Mahrattas, was purported to have diligently sought out these insects to restore his lost vigor in the last century. Caterpillars of the genus *Cossus* were pulverized and prescribed for stimulating lactation in pregnant women possibly because the caterpillar, on being disturbed, secretes a whitish, oily substance.

Perhaps the most imaginative application of the doctrine of signatures involved the use of field thistle galls, thin veinlike excrescences along the plant stem caused by the presence of a minute wasp larva, in the treatment of hemorrhoids. Such powerful medicine was *Carduus hemorrhoidalis,* as it was called by the medieval medical community, that simply carrying a gall on one's person was deemed an effective prophylactic.

The kindest thing to be said for most "sympathetic medicines" is that, while almost assuredly not doing anyone any good, they probably caused no one any lasting damage. But, as such, sympathetic medicines make up only a small portion of entomological materia medica. The vast majority of insect curatives cannot be explained in terms of real or imaginary signs or resemblances. For many remedies, it certainly is the case that insects or insect products have some definite therapeutic value. Botanical medicine, though for the most part as hit-or-miss as insect medicine, has justified itself in the public mind in the form of such indisputably effective drugs as reserpine, aspirin, quinine, opium, steroid hormones, digitalis and the like all of which were first discovered as powerful folk medicines. There is really no objective reason to assume that "bug drugs," in the grand scheme of things, are any less effective on the average than are the botanicals.

As for the qualifications that make insects worthy of consideration in the manufacture of pharmaceuticals, they are by no means wanting. When it comes to the synthesis of chemicals, insects are, as is their wont in most endeavors, tremendously prolific. Not only do they synthesize a wide assortment of their own chemicals defensive sprays, mating and alarm pheromones, venoms and toxins -but many can incorporate into their bodies chemicals from the plants they eat. Suffice it to say that the enormous assortment of chemicals associated with insects includes compounds that are demonstrably emetic, vesicant, irritating, cardioactive, or neurotoxic in short, the insect world is a veritable drugstore of pharmacodynamic agents.

Although medieval physicians did not have at their disposal the technical jargon that doctors today dispense freely, they nonetheless used a bewildering assortment of equally obscure terms and phrases, many of which related to the classification of ailments and restoratives. The system that dominated medicine for more than five centuries was the one proposed by Galen, a physician who lived from A.D. 129 to 200. The publication of his text *De Methode Medendi* so profoundly affected the science of medicine that for centuries afterward certain drugs were referred to as "galenicals." Galen's theory was that good health was the result of a proper balance of the four elements, characterized as hot, cold, wet, and dry. When one of the elements came to predominate over the others, illness resulted. Treatment consisted of the administration of an agent with the opposing element in predominance. Thus, kermes berries, galls of the scale insect *Coccus ilicus,* were possessed of "cooling and drying" i.e., astringent properties, no doubt attributable to their high content of organic acids (up to 60% tannic acid for a related medicinal species, Dyer's coccid, on oak trees). Galls formed by a completely unrelated group of insects, the cynipid gall wasps (such as the Aleppo gall), are similarly rich in tannic acid and are prescribed for many of the same ailments as are the coccid galls. Due to the presence of many free hydroxyl groups, tannic acid and other similar compounds have an extraordinary capacity for binding with proteins a property that may be related to vasoconstriction, or narrowing of the blood vessels. Tannic acid is listed currently both in the British and U.S. pharmacopoeia and is recommended for hemorrhage, gastric bleeding and ulcers, and, until recent times, was produced for commercial use by extraction of various insect galls.

Many insect preparations were deemed "warm and dry" these were prescribed to assist those afflicted with an overabundance of the cold and wet elements catarrhs, listlessness and, most of all, lack of interest or ability in matters sexual, Dr. James writes:

> Ants are warm, dry and aphrodisiac and their acid smell wonderfully enlivens the vital spirits. To use them they must be dissolved with a little salt and the diseased part anointed with the liquor. It is preferable to all sorts of apoplectic and strengthening waters, particularly in the cure of catarrh.

Used for everything from tapeworms to deafness, apoplexy, leprosy, and "the itch," ants won medicinal sanction and official status as *Spiritus Formicarium* (infusion of crushed ants in alcohol) or *Oleum Formicarium* (crushed ants in oil). In infusion, ants were used to eradicate facial hair in children and to counteract various skin disorders such as flora, leprosy, and lentigo (freckles).

Many of these preparations may have appeared to be of use in the treatment of such disorders due to the rubefacient (reddening) properties of formic acid, principal constituent in the sting of many ants. Sprayed on human skin, formic acid in aqueous solution may induce reddening, itching, and irritation. Depending on one's motivational state and personal proclivities, itching and irritation may well have had "aphrodisiac" effects.

Perhaps the most celebrated of topical irritants or aphrodisiacs irrespective of origin is cantharidin, a secretion of the renowned Spanishfly. The principal source of cantharidin is *Lytta vesicatoria*, a brilliant metallic-blue beetle in the family Meloidae (and thus not a fly) that occurs throughout southern Europe eastward to Siberia (and thus not particularly Spanish). The ancient Greeks knew of the ability of the beetle's secretion to raise blisters externally and stimulate the mucosa of the bladder and stomach internally. Pliny reported that people sleeping under a tree full of them suffered from their effects and that merely holding one in the hand was sufficient to stimulate the bladder. Such powerful stuff was, of course, in great demand. Cato of Utica, disposing of royal property at auction, sold off his personal stock for 60,000 sesterces, a considerable sum in those days. Throughout the centuries, known officinally as *Musae Hispanicae* or simply *Cantharides*, the beetle has been prescribed for hydrophobia, rheumatism, gout, wounds, buboes, carbuncles, snakebite, leprosy, catarrh, lice, earache, lethargies and internally for dropsy, and as an emmenagogue, diuretic, and aphrodisiac. By Galen's standards, this insect was "extremely hot."

The efficacy of Spanishfly as a curative was abundantly extolled by one John Buncle, Esquire, after a visit to his physician in 1734:

> *So wonderfully has the great Creator provided for his creature, man; in giving him not only a variety of the most pleasing food, but so fine a medicine, among a thousand other, as the Spanish fly, to save him from the destroying fever and restore him to health again. The benefit is entirely owing to that heating, attenuating, and pungent salt of this fly,*

> *and this fly only, which the divine power and goodness has made a lymphatic purgative, or glandular cathartic for the relief of man, in this fatal and tormenting malady. Vast is our obligation to God for all his providential blessings. Great are the wonders that he doth for the children of men.*

Although John Buncle found no cause for complaint, many of his contemporaries may not have fared so well. Cantharidin, the active principle of *Musae Hispanicae,* is not only "extremely hot," it is also extremely toxic even at low dosages. Taken internally, cantharidin can cause serious gastroenteritis and nephritis; as little as 30 milligrams can be lethal. In fact, in 1698 (some forty years before Buncle penned his panegyric), a London physician named Greenfield was committed to Newgate Prison in a sensational malpractice trial for prescribing an excessive oral dose of *Musae Hispanicae* for a stubborn case of strangury and inadvertently poisoning his patient. While in prison, though, the doctor published a series of papers detailing his more successful experiments with cantharidin in the treatment of bladder and kidney diseases and was eventually released from prison to resume his practice. The reputation of Spanishfly as an aphrodisiac gained considerable notoriety in France in 1772, when the lieutenant général criminél Jean-Pierre Chomel of Marseilles prosecuted the notorious Marquis de Sade for poisoning several prostitutes while attempting to inflame their passions with the covert use of cantharidin.

Cantharidin was an enormously popular medicine in this country until the turn of the century. During the Civil War, when supply ships from Europe were cut off due to naval engagements, physicians resorted to using a native beetle in the same family as the Spanishfly, the "old-fashioned potato beetle," *Epicauta vittata,* so dependent had they become on cantharidin in their practice. Despite the dangers involved in its use, cantharidin as a pure extract still is prescribed, mostly in veterinary medicine, as a rubefacient, counterirritant, and vesicant; it has official recognition in the British, Italian, Swiss, Belgian, and Spanish pharmacopoeias.

Recently, considerable attention has been focused on the natural role of cantharidin in the life of insects that produce it. Its toxicity and distastefulness suggest that it has a defensive function. Indeed, when disturbed, Spanishfly and its relatives in the Meloidae release prodigious amounts of the substance by reflex bleeding at the joints between leg segments (earning them the common name of oil beetles). As a feeding deterrent to insect predators, it is effective at vanishingly low concentrations. The distribution of cantharidin within the body of the beetles suggests yet another function, however. Cantharidin was first isolated and identified in *Lytta vesicatoria* in 1810; Beauregard in 1890 found that in Spanishfly the largest quantities could be obtained from the male seminal vesicles and female ovaries. Biosynthesis seems to be restricted to the male accessory glands. Since males transfer copious amounts of cantharidin to females at mating, the substance may actually serve an aphrodisiac function in the insect.

Yet another beetle, *Dermestes typographicus,* is mentioned by both Dioscorides and Galen for raising blisters, and is recommended by El-Scherif for opening abscesses and furuncles. In the case of *D. typographicus,* otherwise known as the common larder beetle, the ability to raise blisters is not due to the presence of any toxin but by the presence of a thick coat of fine hairs on the larvae that, on contact with sensitive parts of the skin, can cause intense itchiness, redness, and all-around discomfort. Other urticating insects, notably caterpillars of many families, have often been prescribed in the treatment of skin disorders although Thomas Moffett, prominent seventeenth century purveyor of misinformation regarding the natural history of insects and other animals (and father of the little miss who sat on a tuffet eating curds and whey), has to say only that "the Germans know that the hairy Caterpillar dried and powdered stops the flux of the belly and some downy and hairy Caterpillars, by tradition are held to cure children, when they cannot swallow their meat for straightness of their jaws."

By acting on skin and mucous membranes, urticants and especially vesicants may have indeed been of some small service in the treatment of urogenital complaints; there existed, however, a number of remedies for such ailments that may have had at their foundation an entirely different mode of action. Insect blood, or hemolymph, is a remarkable assortment of dissolved inorganic ions, sugars, amino acids and other organic molecules. The precise electrolytic balance of the hemolymph varies tremendously along taxonomic lines. In one group of insects, including mayflies, stoneflies, dragonflies, grasshoppers, and true bugs, the principal electrolytes in the hemolymph are sodium ions and chloride ions (at the other extreme, butterflies, moths, bees, ants, and wasps regulate blood osmolarity principally by varying concentrations of amino acids in solution). These ions, particularly sodium, play an important role in regulating water balance in the bodies of human beings in fact, in the human kidney, the reabsorption of water is in part a function of the sodium concentration of the fluid.

Far and away, the vast majority of insects used in specific cures for bladder and kidney dysfunction are those that have hemolymph characterized by high sodium ion concentrations. Among these is the cicada Dioscorides and Galen both recommended roasted cicadas for bladder pains. Dr. James, in his *Medical Dictionary* of 1745, quotes Schroeder and Dale as remarking, "The ashes of the Cricket (*Gryllus domesticus*) exhibited, are said to be diuretic," and also mentions that "grasshoppers (*Locusta anglica minor vulgatissima*) in a suffumigation relieve under a disury, especially such as is incident to the female sex." Curiously, many of these preparations involve roasting and burning. While such treatment may have made unpleasant medicine somewhat more palatable (or at least less recognizable), it might also be the case that, in the same manner that sodium salts today find use as diuretics and urinary alkalizers (e.g., sodium citrate and sodium acetate), the ancients found certain insects to be a concentrated source of salts and thus prescribed them, in lieu of the relatively precious commodity mined from the earth, for bladder and urinary complaints.

Insect hemolymph is not only rich in inorganic ions, it is also full of antimicrobial secretions of various kinds. In recent years, numerous investigators have amply demonstrated the bactericidal properties of extracts prepared from all manner of insects. Thus, the prescription of insect preparations, particularly insect blood, for amelioration of many kinds of infectious diseases might have had some, even if slight, empirical basis. The bed bug, *Cimex lectularius,* is a case in point. Dioscorides attributed to bed bugs no fewer than eleven medicinal virtues, ranging from external application (mixed with tortoise blood) for wounds to attachment to the left arm "in some wool that has been stolen from the shepherds" to cure nocturnal fevers. In Chinese medicine, bed bugs find their way into almost as many prescriptions. From "Recipes for restoring life":

> *Bedbugs used for the treatment of lip-turning furuncle. Pound seven bedbugs thoroughly with some cooked rice and apply the paste on the lesion. It will effect a quick cure.*

For stinking and gangrenous ulcers, pound some bedbugs with *Shui Lung Ku* [a mixture of oil and lime used for filling cracks in a boat], mix with sesame oil and apply locally.

For stye, drop a little blood squeezed from bedbugs on the affected spot. The swelling will soon subside.

And from *A System of Pharmacopoeia*:

> *In case of chronic ulceration with a gaping wound, apply locally some bedbugs, the heads of which should be removed.*

Not only do many of these prescriptions mention bed bug *blood* specifically, they almost all involve the use of blood to treat forms of bacterial infections or sepsis: stinking ulcers, styes, wounds, and furuncles. That insect blood, or hemolymph, contains antibacterial activity has been known for over 70 years. Virtually all insects are thought to produce a variety of proteins and peptides with pronounced activity against all manner of microbes, including many human pathogens. Thus, the occasional reference to the use of insect blood or headless insects in entomological materia medica, such as Pliny's prescription of headless *Blaps* beetles against king's evil, leprosy, wounds, bruises, and scabs, may represent early recognition of the occurrence of bacteriostatic agents in insect hemolymph.

Perhaps the most scientific use of insect antibiotics arose on the battlefields of war. Harried and underequipped physicians in both the Napoleonic Wars and the American Civil War noticed with alarming frequency that many of the soldiers whom they judged too far gone to save, with deep, gaping, maggot-infested wounds, recovered more quickly and in greater numbers than did the soldiers whom the doctors had elected to treat. During World War I, an American surgeon, W.S. Baer, discovered that maggots of certain species of Diptera, especially those in the genera *Lucilia* and *Wohlfartia,* preferentially consume necrotic tissue and systematically cleanse the wound. Moreover, in the process, the maggots excrete large quantities of allantoin, a nitrogenous material that sterilizes the tissues. For several years, physicians in the U.S. and Canada employed these maggots (raised under stringently aseptic conditions) in cleaning out deep wounds, particularly those associated with open fractures or chronic suppurations; in fact, in 1934, the Entomological Laboratory of the USDA sent out a questionnaire to surgeons who had used the maggots and found that 91.2% of those responding found the method highly satisfactory. With the isolation and synthesis of the antibacterial agent allantoin, followed by the discovery of antibiotics, the practice of maggot therapy has fallen largely into disuse; however, entomology departments around the country still get occasional requests from surgeons for sterile larvae for particularly refractory cases of osteomyelitis.

Ants have also aided physicians by facilitating wound healing but in an entirely different fashion. In the Amazon, army ants were used by native peoples to assist in surgery. Surgical incisions are closed by allowing the sharp serrated mandibles of the soldiers of these species to close around the incision, thus suturing the wound. Once these ants lock their jaws, they remain forever closed, even when the head is severed from the body. Carpenter ants are used in a similar manner in the Mediterranean region, as well as in parts of Africa and India.

One other indirect use of insects to heal was popular around the turn of the twentieth century, before arsenicals (and later antibiotics) were discovered to be effective at destroying the microbe that causes syphilis. Syphilis, a widespread venereal disease, is, if untreated, a progressive illness that leads eventually to debilitation, in some cases insanity, and death. Without effective medicines, physicians were willing to try all kinds of harsh treatments to halt the progression of this disease (including, among other things, drastic regimens of bloodletting and purging). When it was discovered that the causative agent of syphilis (a spirochete) could not survive temperatures over 40°C either outside or inside the body, physicians experimented with deliberately infecting syphilitic patients with malaria by exposing them to the bite of a mosquito carrier. One characteristic of a full-fledged case of malaria is intermittent, very high fever; such fevers were often sufficient to eradicate the syphilis pathogen from the body. Of course, former syphilitics remained infected with malaria for the rest of their lives (at the time, no effective treatment for the malaria pathogen was known), but the consequences of a lifelong infection with one of the mild strains of malaria were vastly preferable to those of tertiary syphilis.

Sticky or slimy insects have often proved medicinal, due to their very sliminess and stickiness, in the formulation of emollient or demulcent plasters. Most cockroaches, for example, are unlike the majority of insects in that their exoskeleton is coated not with a waxy layer but rather with a greasy layer; among other things, the grease allows the roach to slide into narrow cracks and crevices where human feet cannot pursue them. In the Prussian pharmacopoeia, the cockroach rated officinal status (*Pulvis Tarakanae*) in a number of preparations and was employed in Jamaica, bruised and combined with sugar, to treat ulcers. Whole insects were also given to children to eliminate worms, possibly by just lubricating them away.

Insect venoms, used by insects for everything from paralyzing prey to protecting themselves from predation by vertebrates, contain an amazing assortment of powerful proteinaceous materials, many of which can cause severe pain and systemic effects upon injection, a fact to which anyone who has been stung by a wasp or hornet can attest. Scientific investigations of the properties of insect venoms are centuries old. Perhaps one of the earliest investigators was Cleopatra, who, searching for a painless, fast-acting poison, tested wasp venom on condemned prisoners and found it greatly to her liking; bee venom could not be used for such insidious purposes, due to the sacred status of the insect.

The mode of action of bee and wasp venom is thought to involve blockage of sensory nerves and, due to the presence of the enzyme hyaluronidase, increased capillary permeability. Thus, the long-established practice of administering bee stings or bee extracts (known officinally as *Apis*) in the treatment of rheumatism has a possible physiological basis increased capillary permeability means enhanced blood flow to afflicted areas and ganglionic blockage means reduced perception of neuralgic pain. The therapeutic value of bee venom is becoming more widely recognized by contemporary physicians; not only are a number of commercial preparations of bee venom available today (Apicur, Virapin and Forapin, to name a few).

Spider venoms are also showing potential as curatives, particularly for nervous system disorders. Spider venoms are frequently complex mixtures, with over a hundred individual components, and are characterized generally by the exceedingly fine sitespecificity of action of these components. In 1982, one component of the venom of a Japanese spider was found to block the action of glutamate, the most ubiquitous neurotransmitter substance in vertebrate nervous systems (a neurotransmitter substance being a chemical that carries signals from one nerve cell to another).

Specifically, the venom interferes with the passage of ions, or electrically charged atoms, that move in and out of nerve cells to allow them to transmit electrical signals. Because glutamate is thought to be involved in nerve cell death after injury or stroke, spider venoms may be important not only as research tools to be used in elucidating neurological processes but also potentially as curative agents. To supply the newly increased demand for spider venoms from research scientists, several enterprising entrepreneurs have started up businesses devoted to collecting and selling spider venoms; one such firm, located in Black Canyon City, Arizona, is known as Spider Pharm.

The extraordinary toxicity of certain insect defensive compounds has found use throughout the world not in curing disease but in putting food on the table. The people of the Kalahari in Botswana dig up pupae of several species of chrysomelid beetle, including *Diamphidia nigro-ornata,* and apply fluids from these pupae to the shafts of arrows used for hunting. Up to ten pupae may be used to coat a single arrow, which is then dried over fire. Shot at close range, these poison-tipped arrows can bring down game such as giraffe, ostrich, wildebeest, and zebra. The toxin, which is deadly at a dose of 100 to 250 micrograms per kilogram injected into a mouse, causes frantic movement, twitching, and defecation, followed by paralysis and eventually respiratory failure. The active principle has been identified as a protein, dubbed diamphidiatoxin.

Though couched in far more scientific language and divested of its religious overtones, the belief that insects of necessity contain powerful medicaments persists even today in the form of the ongoing investigation of natural products for anticancer activity. Members of the "grind and find" school of investigative pharmacognosy systematically extract and isolate potential cures for cancer from the unlikeliest of places: blue periwinkle flowers, deep-sea sponges, and even insects. To date, these surveys have unearthed a number of highly effective antineoplastic agents from insect sources; cytotoxic and antineoplastic compounds, potentially useful in the treatment of cancer, have been found in the wings of a butterfly, in the legs of a female stag beetle, and in the venom of the common yellow jacket. Although nowadays no one would seriously argue that these potential drugs were placed in insect storage containers to keep until people developed the technology to utilize them, the basic philosophy that there are so many different chemicals produced by insects that at least one of them is going to do something wonderful for people is a sophisticated reaffirmation of the belief that nature has been good to us and a compelling argument for preserving species and habitats in even the most obscure corners of the planet.

Human Entomophagy (Rustling up Some Grubs)

The reaction of most Americans to eating insects is a mixture of distaste and disbelief. The official word for insect-eating, entomophagy, even sounds like some sort of horrible disease. Technically speaking, however, it's a commonplace occurrence in the United States, officially sanctioned by the federal government. The Food and Drug Administration puts out a list of maximum permissible levels of insect infestation or damage, that is, a maximum number of insect adults, eggs, immatures, droppings, or fragments with which a food can be sold. For example, in every 100 grams of Brussels sprouts, there can be no more than 40 aphids and/or thrips; a half-cup of raisins can contain no more than 10 insects or equivalent in pieces and 35 fruit fly eggs. Just about every food imaginable is on the FDA list, and for those interested in tallying their daily insect consumption, you can send away for it for free. For example, there may be as many as 56 insect parts in every peanut butter and jelly sandwich.

The reason these standards exist is, as the FDA brochure states, is that "it is not now possible and never has been possible to grow in the field, harvest, and process some crops that are totally free of natural or unavoidable defects. Even with modern technology all defects in foods cannot be eliminated." The levels of maximum permissible infestation are set not so much for health reasons but for aesthetic reasons; usually the limits are at the lowest level of detectability. Nine leaf miners per 100 grams of spinach, for example, may be no less dangerous than 19 they're just a bit less conspicuous. It has been argued that, since entomophagy is widely practiced worldwide, there should be no legal restrictions on the occurrence of insect parts in food on the contrary, that the label should simply indicate their presence and the consumer be free to decide whether or not to buy the product.

The point of all this is, though, that insect contamination doesn't seem to do anybody any harm (except may be psychologically, after it's been pointed out to them, and physiologically to that proportion of the population with an allergy to insects or insect parts). In fact, insects may be contributing a small but significant fraction of our daily protein intake. Basically, there's nothing inedible about insects from the human perspective. A quick examination of the composition of insects reveals that they're not so different from beef, pork, or fish. The chitinous exoskeleton is by and large indigestible, but then again, so is apple skin, and the exoskeleton makes up only a small part of the total biomass (about 4% in caterpillars) and doesn't affect the nutritive value of the insects as food. Analysis of the protein content reveals that insects contain all the essential amino acids (those required by humans for synthesizing proteins), and so are superior as a protein source to most plant proteins. Most insects have an overall higher protein content than do plant foods crude protein content of caterpillars ranges from 30 to 80%. Insects are even rich in vitamins and minerals (so maybe it's the worm a day and not the apple it's eating that keeps the doctor away).

Not only are insects nutritionally suitable as a food source, they're economically feasible as well. An important consideration in raising livestock is a measure called the efficiency of conversion of ingested food (ECI), the amount of weight an animal gains per weight of food an animal eats. Chickens are very good at converting the food they eat into food we can eat; for every 100 pounds of chicken feed, 38-40 pounds of chicken meat are produced. Beef cattle produce only about 10 pounds of meat for every 100 pounds consumed (this is one reason that chicken is cheaper than beef). Another way of looking at this ratio is that 90% of what a steer eats is wasted as far as the consumer is concerned (though probably not from the steer's perspective). ECI measurements for insects are not all that easy to make nor are the ones that have been made completely accurate (they vary with food plant, condition of food plant, age, sex, and temperature, among other things, and there's a lot of error involved in weighing caterpillar droppings). On the average, however, insect ECI values are quite respectable. Caterpillars such as silkworm and tent caterpillar, with ECI values of 19-31 and 14-22 respectively, put sheep, with a paltry 5.3, to shame. What is also important is that insects can be fed food that humans cannot consume directly, such as mulberry leaves and cherry tree leaves; cows and sheep, which are often fed grains such as corn and wheat, are in a sense competing with people for food.

Insect-harvesting may actually be a more energy-efficient way to accumulate protein and calories than hunting for game. David Madsen collected sun-dried migratory grasshoppers (*Melanoplus sanguinipes*) that accumulate on the shores of the Great Salt Lake in Utah in enormous numbers (and which were used historically by many native Great Basin inhabitants). Wave action had deposited numbers of grasshoppers in rows ranging upwards of 6 feet wide and 9 inches thick (up to 10,000 grasshoppers per foot). He found that it was feasible for one person to collect some 200 pounds of grasshoppers in an hour; since the grasshoppers contained approximately 1,365 calories per pound (versus 1,240 per pound for beef), a single gatherer could conceivably accumulate some 273,000 calories per hour of effort invested (dried and salted, the grasshoppers also would require no elaborate cooking or preserving procedures). In contrast, hunting for big game returns only about 25,000 calories per hour of effort invested. While such literal windfalls of insect protein are undoubtedly unusual, Madsen and his colleagues managed to collect 18.5 pounds of crickets in an hour simply by gathering them up from the bushes. Since dried crickets contain about 1,270 calories/pound, their hour of work yielded the caloric equivalent of forty-three McDonald's Big Mac sandwiches.

Not only does reason dictate in favor of entomophagy, so, it appears, does instinct. Insect-eating has been a common theme throughout human evolution. Humans may have been retained the habit from simian forebears; many primates regularly consume insects as part of their diet and some, like the Malayan spectral lemur, exist almost exclusively on insects. Chimpanzees use twigs as tools for extracting termites from their mounds, and such activities can make up over 30% of their foraging time. Gorillas, too, have been observed deliberately partaking of insect (particularly ant and termite) meals. Moreover, there is some evidence that certain species of primates actually produce chitinase, an enzyme that would allow them to digest the exoskeleton.

Worldwide, the most commonly consumed insects are those that are available in large quantities specifically, migratory locusts or social species such as termites. In virtually all entomophagous cultures, however, certain rare species are consumed not so much for their nutritive value but for their flavor. Many true bugs, which produce pungent secretions from defensive glands, are used in condiments; honeydew-producing homopterans are valued for their sweetness. The earliest known human records documenting entomophagy as a regular practice (aside from cave paintings depicting honey hunts) is a bas-relief executed about 700 B.C. showing servants of King Sennacherib of Assyria carrying locusts skewered on sticks in preparation for a feast. Near Ninevah, in the palace of Asurbanipal there is a contemporary bas-relief depicting a similar scene. The Old Testament makes recommendations as to which insects should or shouldn't be eaten. From Leviticus 11 (21-23): "You may eat of all winged creeping things that go upon all fours which have legs above their feet, wherewith to leap upon the earth.

These you may eat; the *arbeh* after his kind, the *sal'am* after his kind, the *chargol* after his kind, and the *chagav* after his kind" (all names of various developmental stages of locusts scholars love to dispute this point). The Mishnah Torah, a collection of commentaries on the Bible, interprets the passage to mean "among locusts, these are clean; all that have four legs, four wings, and jointed legs, and whose wings cover the greater part of the body." In the Koran, the prophet Mohammed is said to have eaten locusts while discoursing with his disciples. Locusts are permissible food according to Moslem dietary laws. Mohammed states, "Lawful for us are two dead (animals) and two blood liver and spleen, fish and locusts." While locusts may have been technically permissible, they were not considered choice fare: there's an old Arabic proverb that goes, "A date is better than a locust." The ancient Greeks were avid insect eaters, too. One of the earliest Greek references to entomophagy comes from a work by no less a personage than Aristotle of Athens. In his *Historia Animalium* he reports, "the larva of the cicada on attaining full size in the ground becomes a nymph; then it tastes best, before the husk is broken as for adults at first the males are better to eat after copulation than the females, which are then full of white eggs." Grasshoppers of various descriptions were sold in the marketplaces to be ground into flour for cakes or simply served as hors d'oeuvres. Aristophanes refers to "four-winged fowl" at the markets for the poorer classes. Greek historians tell of a people in Ethiopia whom they called the Acridophagi the locust-eaters (Diodorus, second century A.D., remarked that they were a shortlived sickly bunch).

The Romans, with typical Roman exuberance, indulged in a few unusual insect delicacies. Chief among these was the cossus, a grub or caterpillar of uncertain identity that in Rome was considered a delicacy after it had been fed with flour. Under more natural circumstances they bore into the trunks of oak trees. After the Romans, though, insect-eating among Westerners was largely a matter of fending off starvation (as was the case during a few European locust plagues) or of personal eccentricity. It's hard to say why and how western culture lost its appetite for insects. Perhaps more energetically effective methods for raising domestic livestock supplanted the need for supplementary insect protein.

On other continents, however, insect-eating was or is a fact of life. Many of the cultures that eat insects, living in hostile and unproductive environments, even rely upon insects as their major protein source. In many preagricultural societies, fast disappearing under the encroachment of Western agricultural practices, hunting and gathering are the means by which food is secured, and nothing even remotely edible is overlooked. Plagues of locusts never caused the despair in Africa that they have in Europe for the simple reason that, in the absence of their normal food plants, people of the African bush have historically just as happily eaten the locusts.

"Bushmen's rice" is actually a kind of ant with a long body and black head; Hottentots like them, too. The Pedi of South Africa, the Shona of Rhodesia, and the Hottentots of South Africa are locust lovers, as are the Bushmen, and rise before dawn to snag the insects before they warm up and begin moving about. The locusts are customarily boiled, cooled, dried, cleaned, and salted; the legs are ground into flour and cooked with salt and peanut butter. Termites are a popular item roasted, dried, or fried in fat throughout Africa, as is the soil from termite mounds. Grubs that tunnel in the trunks of palm trees are eagerly sought and are located by leaning against the tree and listening for them. In Cameroon, the Pange eat a number of insect foods, including no fewer than twenty-one kinds of caterpillars. Almost three dozen species of caterpillars are consumed in the Shaba region of Zaire, and, in the Kwango District, the production of dried caterpillars for consumption exceeds 280 tons per year. In South Africa, roasted mopane worms (the immature stage of a large moth) are bagged and sold as snacks, much as cheese puffs are in the

United States.

Silkworm larvae and pupae even today can be purchased in cans in supermarkets. Ricehoppers are served cooked in soy sauce and sugar (with legs and wings removed) and are so popular that they are processed and sold in supermarkets to be served as cocktail snacks (called *tsukudani*). According to the National Food Composition Tables, compiled by the Resources Council of Japan, *tsukudani* contain 22.5% protein (and are probably more nutritious than beer nuts). Mass production of ricehoppers is centered in Nagano prefecture, where insect eating is very popular at least ten orders are consumed on a regular basis. In the southern part of the prefecture, insects may contribute in an important way to the mineral nutrition of the inhabitants. Nanshin district, for example, is surrounded by mountains and is subject to heavy rains, making transport of salt into the area difficult. Insects may thus be an important source of sodium and other minerals in the diet.

The aboriginal inhabitants of the regions to the east were, due to favorable climatic conditions, agriculturists early on. Entomophagy was most commonly practiced in the arid Southwest and Mexico. Rocky Mountain locusts, which underwent periodic population explosions and destroyed everything green for thousands of acres prior to their extinction in the early part of the twentieth century, were eaten with great gusto by the Assiniboine and Shoshone. These Native Americans used to go on locust drives, like cattle roundups, in which they beat through the bushes and drove the locusts into an open pit. The Paiutes of the southwestern U.S. ate certain species of pine-feeding caterpillars, which they smoked out trees by building small brush fires under the canopy, thus capturing and cooking them simultaneously. The Digger Indians of California caught ants by spreading damp skin or bark over anthills and collecting what accumulated there. The Cheyennes, the Snakes, and the Utes ate lice, ants, grasshoppers, and cicadas. Brine flies, which also undergo periodic population explosions and mass in enormous numbers on lakeshores, were favorites of many tribes in California, Utah, and Nevada.

The use of insects as foods in Mexico (although primarily as condiments or special treats rather than as staples) continues today. Agave caterpillars (*gusanos de maguey?*) of several species are served as appetizers or as filling ingredients for tostadas or tacos. These same insects can be found floating in the bottom of bottles of mescal, a liquor brewed from agave; the presence of the worm is presumably a guarantee of authenticity, since the caterpillars can be found only in agave plants. *Jumiles,* or stink bugs, are used to season chili, and eggs of waterboatmen (*abuahutl*) are ground into flour for cakes.

Throughout the history of Western civilization there have been but few enlightened individuals who have promoted the use of insects for culinary purposes. Linnaeus, the celebrated taxonomist, recognized that such practices existed in fact, of the palmworm weevil grub he wrote "*larvae assatae in deliciis habentur*" ("roasted or broiled larvae are considered a favorite food"). Ludolph cooked the locusts that were plaguing Germany in 1693 like crayfish, pickled them in vinegar and pepper, and served them to the Council of Frankfurt, summoned to discuss the plague problem. The notorious Barbary pirates included locusts as regular items on their bill of fare and David Livingstone, of "I presume" fame, ate locusts in Botswana when other food supplies ran short. C.V. Riley, head of the U.S. Entomological Commission at the time of the great locust plague of 1873-76, prepared a number of locust recipes and in fact served an entire locust dinner (soup to nuts) to members of Congress to dramatize his feelings that insect-eating might alleviate much of the suffering brought about by the insects. The great Charles Darwin himself, in his book *Phytologia*, commented on the Chinese habit of entornophagy and mentioned that he himself had sampled some caterpillars of a culinary variety (a sphingid, or hornworm) and found them delicious.

While the advocates of entomophagy per se are few, there are more people interested in using insects as a source of protein for other food animals. The ability of many insect species to develop on material that is otherwise unusable is an additional benefit (e.g., farming chironomid midge larvae on manure to feed to shrimp not only produces shrimp but cleans up polluted areas). The negative psychological impact of consuming insects is greatly attenuated when it is one link removed on the food chain, too. According to extensive studies, people cannot detect the difference between meals prepared with chicken raised on chicken feed and those prepared with chickens raised on Mormon crickets.

There is of late increasing interest, particularly in third world countries, in exploiting insect protein commercially. In South Korea, for example, some rice farmers no longer spray insecticide to kill the rice-field grasshoppers known locally as *metdugi* but instead harvest them and sell them in addition to or instead of the rice itself; in Mexico, scientists at the National Autonomous University have been developing methods to increase yields of agave worms, a cherished and costly delicacy not only in Mexico but in Japan as well. Dr. Gene Defoliart, at the University of Wisconsin at Madison, founded a periodical, dubbed the *Food Insect Newsletter*, to promulgate scientific information on the history, culture, and feasibility of entornophagy. Other pockets of interest exist as well. One case in point is the Insect Club--a nightclub/bar/ restaurant in northwest Washington, D.C., that specializes in, along with insect decor, insect menu items such as mealworm wontons, cricket peanut butter cups, and cricket brittle with ginger essence, prepared with insects reared pesticide-free and shipped fresh regularly from California. At present, however, unless American attitudes undergo a shift of unparalleled proportions, such establishments are likely to remain interesting oddities, frequented perhaps by people who have accepted a dare or had the misfortune to lose a bet.

10

Virtue and Practice of Bioethics in Life Sciences

Introduction

We describe our goals, certain virtues are necessary to accomplish them. When I asked people what virtues they thought were important for the work, the most common answers referred to open-mindedness: "flexibility," "the capacity to have multiple perspectives, to listen carefully." My question about what vices, if any, particularly impair good work was often answered correlatively: "rigidity," "dogma." Many of the people with whom I spoke were thinking in terms of individual encounters, but Larry Churchill has made the same point more globally. He argued that we must maintain "skepticism about our usefulness, agnosticism about our answers, pragmatism about our usefulness, empiricism in our endeavor, and irony in our use of 'expert'" together with utter seriousness about the task.

One could call each of those attributes a virtue. In naming virtues, one need not stick to a classical vocabulary: one might instead include something like adventurousness, for instance. One researcher found those who rescued Jews during the Holocaust to be generally adventurous. Nor need we draw only from the Western tradition: Buddhism has interesting things to say about compassion, and the

Confucian scholar Mencius thought about emotions in quite a different way than Aristotle did. (Emotions are partly constitutive of virtue, a virtue being a disposition to act in the right way for the right reasons.) Even within our own tradition we need not strain to find a conceptual rigor that is not there. Zag zebski points out that "our language does not contain a sufficient number of names that convey the full unified reality of each virtue. Some names pick out reactive feelings (empathy), some pick out desires (curiosity), some pick out motivations to act (benevolence), whereas others pick out patterns of acting that appear to be independent of feeling and motive (fairness). For this reason it is easy to confuse a virtue with a feeling in some cases (empathy, compassion), and with a skill in others (fairness). The result is that it is very difficult to give a unitary account of virtues using common virtue language."

Taking advantage of the freedom this description offers, I will speak rather loosely of some virtues that I believe are required in bioethics. Some of my points could be made under different rubrics; the topic of this chapter, about taking on the right projects, might be approached as a matter of role obligation. I stay with the virtue model because it naturally leads us to think in terms of whole persons and whole lives, rather than specific actions. And of course a focus on virtue follows naturally from considering bioethics as a practice; MacIntyre's major purpose in introducing the concept was to provide a contemporary way of thinking about virtue.

Many different virtues are required of the bioethicist; there is no algorithm for producing a list. Many of us teach, an activity that demands humility, courage, and a sort of cheerful selflessness. In all of our activities, we need to listen, and Joseph Beatty has argued that the ability to do so well is a virtue of major significance. The people with whom I spoke offered openness, compassion, fairness, tact, truthfulness, and "a certain idealism, a willingness to deal with the issues society doesn't want to: either go deeper, or actually take a stand." For his part, MacIntyre would argue that honesty, courage, and justice are necessary for bioethics as they are for all practices, and I do not disagree.

My own discussion will focus on choosing the right projects (which I consider a virtue in itself), and the way internal and external goods place pressure on the effort to do so; on the moral demands of interdisciplinary work; and on what it means to be a virtuous community, one in which the goals of the practice are likely to be accomplished. Along the way I will discuss some traditional virtues as well, especially integrity, discernment, humility, and courage.

Choosing Projects Well

It matters how we spend our time. One professional with whom I talked considered this a central challenge in bioethics. Asked to name a significant virtue, he offered these thoughts: "Courage; or whatever it takes to be credible; not so much private life morality, but things like passion, conviction. On the other hand, there are missionaries in the pejorative sense, too, imposing their values on others. Fine line. But [we] should be proactive, not just responding to the questions others put to us. The real question is, `How do you do ethics?' And the answer is, `From the big picture, from the point of view of the marginalized, from the fact that we exploit people.'"

I think that he is right about the way we should be choosing our projects, and right about the fact that it takes courage. It also demands self-awareness. One of my interviewees was asked to do a series of talks to a church group, whose pastor had taken a bioethics course elsewhere. There were to be three sessions under the general title "When Someone in Your Family Is Ill, Aging, or Dying." The pastor wanted the section on family illness to deal with anencephalic infants, an issue that had fascinated his teacher and so fascinated him. But the woman with whom I spoke thought a discussion of chronic illness cancer or diabetes would much better meet the needs of that audience. She believed the pastor had been ill-served by a bioethics course that left him insensitive to the needs of his congregation. Someone else remarked, "I sat on a committee for *ten years* without knowing that pain was undertreated. This is a Catholic hospital—it shouldn't be a problem. But no one ever told us that hospital ethics committees should pay attention to it."

The forces that shape our professional agendas deserve attention. During the week of November 7, 1994, I was struck by how much expensive professional time went into the discussion of Mrs. Bactri's refusal of a gastrostomy. I came to realize that what was at stake was not how much longer she would live, but how decisions made at the end of her life would shape its meaning. I still remember that session as rewarding, in spite of my doubts in principle about ethics consultations. Upholding Mrs. Bactri's refusal of treatment was an honorable and significant activity, and still more so if we contributed to a pattern of respect for patient choice within the hospital.

But it would be hard to argue that this was the only, or the most significant, thing that any of us could have done that day. As I realized toward the end of the week, improving decision making within one area of medicine may divert attention from others. Attention is a scarce commodity, and it does not automatically go to what matters most. We know, for instance, that public concern with drugs, crime, and other issues varies not with the current magnitude of the problem but with the degree of attention the media have given it. That is one of the reasons our interactions with the press are important. But we are catalysts of attention in everything else we do as well, in a private ethics consultation or on a web site visited by millions, in the classroom or presenting Grand Rounds. I don't mean, of course, that we are constantly followed by avid fans. Many ethics committees within hospitals are virtually invisible. I simply mean that, since part of our work is getting people to pay attention and think, we need to be self-critical about where we direct our energies.

Choosing well in this regard may be thought of as a virtue. And to choose well, we must have a rather sophisticated understanding of not only what is important within our domain, but also the forces that move us toward certain questions, and the feasibility of trying to resist those forces. (Audiences cannot just be told, "This matters Pay attention.") This broad understanding demands a disciplined, habitual attention and openness. In addition, choosing well demands courage. Asking new questions can put one at risk, not personally, but professionally. It can demand self-sacrifice: ignoring the possibility of grant money, or surrendering old easy ways of captivating an audience (whether undergraduates or the dean of the medical school). If "choosing projects well" involves moral understanding, a shaping of emotion and desire, and habitual action, then it's a reasonable candidate for a virtue.

The virtue would demand, among other things, an understanding of the currents in the ocean in which we swim. Why do we find ourselves in some parts of the sea and not in others? It is natural to assume that we are moved by the intrinsic gravity of certain questions those involving death, in particular. Casting some doubt upon that assumption is an important first step.

Focus: Death

For many years the medical school ethics curricula at MSU focused almost exclusively on death (or what is referred to within the field as "end-of-life" issues). We were not alone among medical schools in that preoccupation. Brain death, persistent vegetative state, the right of competent patients to refuse life-saving treatment, euthanasia, physician-assisted suicide, treating or not treating severely compromised newborns: ten weeks of this, ten different facets of the same fundamental question when is it morally permissible to take a course of action whose foreseeable result is death?

The question is hardly unimportant. Nor is it a single, simple question: there are many "end-of-life" issues, each raising separate, deep questions. Respecting a refusal of treatment involves questions of patient rights and of what counts as decision-making capacity. Parents' refusal of treatment for a newborn raises very different issues, about their rights over their children and the moral status of infants. And so on; one could spend a productive year-long graduate seminar on any of these issues.

But they are not the only ethical questions that arise in health care, nor are they the most common. Life is saturated with moral questions, assumptions, and challenges, within health care as well as outside it. If morality is a question of properly valuing what is and what could be, then everything we do has a moral dimension. Fortunately that dimension does not need constant attention. We can and must take a great deal for granted, morally as well as practically, but there is endless material for reflection, should anyone look for it. Later I will talk in detail about moral problems in health care that get relatively little bioethics attention. For the moment a few initial examples might help: What does a nurse in intensive care say to telephone inquiries from patients' friends, relatives, neighbors? What kind of hospital advertising is legitimate? How much time should doctors devote to charity care? How, and how often, should alcoholics be treated? Should lab technicians unionize? Should Medicaid patients be shifted to managed care?

One natural response to this list is to say, "Yes. These things matter. But nothing matters as much as death." In one sense, of course, that is true; death is uniquely important. But it does not follow that every *decision* about death is uniquely important. Some decisions to allow death really are profound, but others are not, and they do not exhaust the domain of what matters. Where years of life are at stake (especially years of conscious life), choosing to allow death is a grave matter. Most of the cases that have reached the courts or served as the impetus for legislation and thus shaped bioethics concerned this kind of decision. Karen Ann Quinlan and Nancy Cruzan were in persistent vegetative states, and could have lived for years that way. (In fact, Quinlan did; it turned out that she did not need the ventilator support whose withdrawal the court allowed.) "Baby Doe" in Bloomington, Indiana, who had Down's syndrome and was allowed to die from a correctable intestinal blockage, would probably have lived well into adulthood. Dax Cowart, the badly burned man treated against his will, is still alive decades later.

Most questions about allowing death are different. Death today is usually "managed": that is, medical care delays it somewhat, and then discontinuing care allows it. In my experience, ethics consultations are about this kind of decision more often than any other. In one extreme instance a consultation was requested about an unconscious patient with no family, who would have lived a few weeks if given aggressive care, and a few days without it, in neither case regaining consciousness. In another case a baby would have died within a few days with one kind of care, within a few hours without it. Often what was at stake in these cases was the question of who had the right, or should be forced to shoulder the responsibility, of deciding whether to try to extend a severely compromised life for a short time.

This question is not an inevitable focus for bioethics attention. It is too easy to imagine, and in fact to remember, a system in which these decisions were routinized: some decisions automatically fell to doctors, others to patient or family. Even today, many other kinds of decisions are essentially made for patients: "This is what we're going to do." "The leg has to come off." "I'm going to change your medication." (For some of these actions, the patient will be asked to consent, but only as a legal formality. A really resistant patient could say no, but that is generally seen as a practical or psychological problem, not an ethics issue.) I am neither endorsing nor criticizing what is routinely done, but am simply pointing out that it is not the intrinsic gravity of an outcome, but the attention we have learned to pay to *choices* about it, that creates a bioethics frame. Bioethics is often not responding to something naturally compelling, but to something it has helped construct as compelling. That this has happened may be right and good, or it may not; but it is not inevitable.

Factors That Direct Our Attention

If a focus on (choices about) death is not inevitable, then probably no focus is. What are the factors that shape our choice of projects? If we are to choose them well, we need to understand what there is to resist, and what to endorse.

Impact of History

What we call bioethics arose in the 1960s and 1970s in tandem with movements for civil and consumer rights; medical paternalism and research on unconsenting subjects were seen as violating the same basic rights to information, self-determination, and choice. About the same time new technologies kidney dialysis, organ transplants, in vitro fertilization, intensive-care units were forcing hard choices. Within this context novel litigation occurred: Severely compromised lives, for instance, could now be prolonged indefinitely, and sometimes these patients or their families went to court to have life-sustaining treatment discontinued. This history of bioethics has been traced by others in detail, and like any history it includes, besides these broad cultural forces, personalities, power struggles, and historical accidents. For the first few decades the story was largely an American one, and as the field develops today on other continents it retains an American imprint.

The results of those early activities still inform the field: national commissions produced reports which remain authoritative; judicial decisions set precedents; federal regulations about research on human and animal subjects still have as their core what was formulated then. A body of scholarly literature set an intellectual agenda; hospital ethics committees and the custom of consultations about patient care began; undergraduate courses were introduced, attracting from the beginning large enrollments (which every philosophy department needed, and still needs). All of these undertakings helped the field continue, and they continue to shape it.

And so a certain set of issues remains central today: "patients' rights," which is most strongly a right to refuse treatment; "informed consent" by subjects of scientific research; new technologies and the questions they pose, currently those arising from genetic science. I would not argue that we are free to ignore these historical roots. Those who in various senses employ us expect us to address such issues, and their expectations create both moral demands and practical limits. But just as the Supreme Court finds within its own precedents the seeds of different decisions for different times, we need to look within our traditional concerns for fundamental principles that will allow us greater scope.

Economic Forces

Another pattern directing the attention of bioethics is economic; we attend to issues that interest those who, in a variety of ways, pay us. The most basic form of payment is a salary and a secure job. My own center began, in part, because its early nonphysician faculty were willing to work on a temporary, part-time basis. (Nonacademic readers, or academics who have always had full-time jobs, may have no idea how poorly part-time faculty in the humanities are paid. It is nothing like pro rata. It can be less than unskilled labor makes at the same institution; and of course it carries no benefits.) In such a context, how do full-time jobs develop? By convincing the medical school that one's work is important; and the best way to do that is to look at questions that trouble physicians, and to look at them from their point of view. If, on the other hand, one is hired into a college of arts and letters because courses in bioethics are in great demand, we are well advised to cover the material that students find compelling. None of this is bad; the vaunted intellectual freedom provided by tenure is always tempered by such forces. But one result is that we tend to put our energies into what has *already* gripped our audiences.

Another almost equally important economic force is the need to get funding from external agencies: foundations, government agencies, and industry. Attention is now being paid to the way funding from industry affects what gets published, and therefore the information available to other scientists: such funding affects the way questions are framed, data are interpreted, and results chosen for publication. Not a lot of attention goes yet to the way funding determines which questions are asked. When the Human Genome Project began, one of the biggest projects in biological science ever launched, 5 percent of its budget was earmarked for the study of "Ethical, Legal, and Social Implications" (ELSI). This was the largest sum of money ever made available in bioethics, and immediately we began to pursue it. Repeatedly, when I asked my interviewees with "genethics" projects how they got started, the answer was, essentially, "Someone told us about a big pot of money up in Washington, D.C. We decided to get some." This should not be written off as opportunism or venality; many bioethicists either depend on grant money for their own salaries, or work within academic institutions where grant getting is virtually demanded.

The ways in which external funding shapes academic effort need considerably more attention. The picture is complex. To begin with, most scientific research is expensive. It needs high-priced equipment (labs, computers, materials) and a sizable support staff. No university can support such projects on its own. In the United States, the largest single source is federal funding, especially from the National Science Foundation (NSF) and the National Institutes of Health (NIH), but private philanthropy can also be significant, especially for the social sciences. In addition, industry helps fund projects with commercial promise. Each of these sources is influenced by a variety of factors. Federal funds are targeted in accord with congressional mandate; with initiatives established by agency staff; and, since decisions about particular projects are made by panels of nongovernmental researchers, with the community of science. When a scholarly field is in its infancy, the staff of an agency have more leeway; one of my interviewees from engineering ethics described the beginnings of this field as partly an initiative from NSF staff, who were much less constrained than counterparts working with, say, the history or philosophy of science, where fields and boundaries were firmly established.

Private philanthropy is shaped by the vision of their founders and, again, by staff. (The Bill and Melinda Gates Foundation is now the largest in the world. Gates is discovering that giving money away usefully takes as much time, energy, and creativity as earning it. His foundation is run by a former Microsoft executive and by Gates's father; some critics have advised him to trade them for more experienced people from older foundations.) One of the people with whom I talked described how a private foundation worried about overpopulation: "They tried to move the conversation from 'population crisis' to 'global stewardship,' then realized the move would be a 'conversion experience' and those are religious terms. The conversion experience could be facilitated by churches. But they had failed to get religious groups on board from the beginning, and recognized that they needed help." The result was substantial grant money made available to a faith-based institution.

This intricate web of forces would seem to have little effect in bioethics, which does not generally require labs, equipment, and staff. But the picture for science influences that for bioethics. Medical schools need money and want prestige, and both come in significant measure from getting grants to do scientific research. The result can be pressure on bioethics faculty to get grants, sometimes quite unpleasant pressure. Someone who had left a medical school for a more traditional academic campus remembered how glad he had been to leave. "The exact same day I accepted this job my dean had raked me over the coals for not bringing in enough grant money. `Where's the good-faith effort, friend?"'

Grant funding, in fact, is highly prized in any department on campus, because grants usually include an item called "indirect costs" money to compensate the institution for its overhead. This can be a high fraction of the money awarded to the project. If, for instance, half a million dollars is awarded for some project, the grant will probably carry *another $200,000* to be given directly to the school or hospital, to be put into its operating fund and used in whatever way it chooses. Those who bring in such amounts are prized not only for the research they do, but for the way in which they help the institution survive. Bioethicists working in a medical school will seem like parasites unless they, too, bring in money. This sense is intensified by the facts that most of us are not clinicians (and so cannot bring in patient-care dollars) and teach relatively little. A medical school might require its faculty to spend about 35 hours a year in the classroom, often with a group of 8 to 10 students; their counterparts in philosophy or literature will spend about 200 hours a year, in the classroom with 10 to 300 students. Finally, the few bioethicists who work in self-supporting centers may depend completely on funding from outside sources sometimes from consulting fees, but almost always from grant agencies as well.

The pressure to bring in grant money, then, is understandable. But it has serious effects on how we spend our time. One person remembered unhappily, "I suddenly found that I was committed to spend the next three summers working on informed consent for tonsillectomies because that was what we had grant money for." On her own she would have chosen to work on questions she found more substantive. The pressure to pursue grant funding presents steady moral challenges: should this person, for instance, have withdrawn from the project she found trivial? In the next chapter I will look more directly at such issues.

Money speaks in a more minor but sometimes cumulatively important way through speaker fees; some of us earn substantial amounts for speaking to lay and professional audiences. A popular and effective speaker, in fact, can make a great deal of money. Audiences are eager to hear about the dazzling issues, once surrogate motherhood, then physician-assisted suicide, most recently cloning.

Finally, there are metaphorical payments. Journalists do not pay us, but media coverage is flattering. Church and civic groups usually offer only modest stipends at best, but reward us with gratitude and attention. The pressure here, again, is to come up with exciting cases and issues, whether or not they are particularly important.

In many ways these audiences, from church groups and students through medical schools and funding agencies, with their varying powers to compensate us, converge in their interests: each is likely to see the interesting questions in health care as those faced by doctors or raised by scientists. Students and other audiences quite naturally like exciting issues; referees are likely to be conservative, to keep within a field's established boundaries. And so we are reinforced in our attention to some topics and not to others. The virtue of choosing projects well demands recognizing the power of these rewards and considering whether and how to resist them.

Academic Factors

"Garbage is garbage, but the history of garbage is scholarship." I don't remember who made this point, but it's true, and not as cynical as it sounds. Furthermore, once a history of garbage is written, there are likely to be responses, criticisms, responses to criticisms, criticisms of responses. Attention begets attention; scholarship generates scholarship. Soon there's a whole field. "Orphan" pieces, unrelated to what others have been talking and writing about, have a hard time finding an audience.

The rhetoric of scholarly fields also perpetuates certain emphases. One example is the conventional point of view in descriptions of ethics cases. Tod Chambers has pointed out one typical way of reporting the cases that are a mainstay of bioethics: the reports "do not tell the patient's story, nor do they tell the ethicist's story; instead they tell the physician's story." Frequently a case study suggests "the clinical gaze. First it identifies the patient by his initials," age, and sex; uses phrases like "admitted to" one problem, "denied" another, "reported" another. Such sentences, written in a passive third person, are nevertheless not neutral. The patient could not tell the story that way. A patient who said "I admitted to no sleep disturbance" or "I reported no childhood diseases" would be suggesting something quite different from what is intended by the conventional "he admitted/denied/reported."

So case studies, whatever their grammatical person and voice, nevertheless present the world from a specific perspective, usually that of the doctor or nurse. In addition there are code words that encapsulate whole ethical discussions: "informed consent" suggests not a process of shared decision making but a patient who listens and says yes; "maternal-fetal conflict" frames a conflict between a pregnant woman and her obstetrician as a choice the latter must make between respecting the woman and protecting her fetus. "Confidentiality" ignores a vast web of questions about who should know what, focusing narrowly on one: whether a health-care provider may reveal patient information to a third party. These technical phrases, these code words, are the tools of our trade. Bibliographies and search engines are arranged around them. Even my own file cabinets are organized accordingly. I have "oncology" but not cancer, "newborns" rather than infants (because bioethics focuses on the questions that are evident in the first days of life, not those that arise during, say, the first year of life). In fact, the only reason I still have a file called newborns is that "neonates" is spelled so similarly that in looking for the latter, I find the former. I do have both "aging" and "geriatrics," however. The virtue of choosing projects well requires noticing such things, holding on to labels like "childbirth" rather than "obstetrics, " habitually asking whether what is included within standard labels could be expanded, and struggling to widen or divert the happenstance currents of scholarly fashion.

Through a Doctor's Eyes

I asked someone who works broadly in professional ethics why he takes up the topics he does. "They're what bother practitioners. Truth telling in the courtroom bothers lawyers. Engineers are haunted by stories of buildings that fall down, deaths and injuries that follow." And so are doctors, of course, haunted by death.

In addition, in a doctor's daily professional life, as one commented, "Refusal of treatment really gets our attention." Such refusals are disruptive rather than haunting, although the refusal of life-sustaining treatment is both, and this fact goes a long way to explain bioethics' focus on death. Arguably a first responsibility of anyone in professional ethics is to do exactly this, deal with what is most troubling to members of the profession. If it's first, however, it is not also last. There is an instructive contrast in the academic field of criminal justice. Researchers, investigating the extreme unreliability of eyewitness identifications, developed methods of conducting police line-ups that substantially reduce the number of misidentifications, *even though lawyers remain uninterested in the results and make no use of them*. It helps that these researchers were tenured academics, not dependent on lawyers for their living or their status.

In bioethics most of us are not so independent, and this makes it harder for us to do what someone whom I quoted earlier demands of himself: to remember the big picture, the perspectives of the less powerful. Take, for example, the fact that doctors and nurses can make problems for one another; either can interfere with good patient care or be personally diffcult. But it is only nurses who identify such problems as ethical issues. Their ethics textbooks often have a section on doctor-nurse conflicts. I've never seen a textbook, or even a syllabus, for medical students that mentions the issue. The reason is that the diffculties are not symmetrical: nurses have real power but little authority, while doctors have both. When nurses get in the way of doctors' doing their jobs, the problem is largely seen as practical. Can the doctor make the nurse see reason, or remove him from the case, or should she just put up with the unpleasantness? For nurses, however, the problem can be a moral one. Because they have less authority to work on their own, they may feel forced to take part in care they believe is inappropriate, dangerous, or ethically wrong. Finding a way not to do that takes both courage and discernment. The asymmetry in power leads to an asymmetry in moral situation, which few doctors recognize.

So it's not surprising that what troubles doctors tends to be decisions about individual cases. When they pay attention to the ethical aspects of organizations and systems, it is ordinarily because something interferes with the care of individual patients, or because the doctors believe they themselves are being mistreated in some way. The massive attention that went toward issues in managed care in the middle 1990s arose first because *doctors* were frustrated. They had been trained to do everything possible for their patients as individuals, not to consider whether using money for this patient would end up hurting others. The public reaction to managed care began slightly later, but it took much the same course: middle-class "paytients" (as Len Fleck puts it) were not accustomed to restrictions and limitations. Of course some of the changes brought by managed care were unwise and unethical, even if some were necessary and fair. My point is that attention to systems entered bioethics at the point when doctors began to find systems seriously troublesome. Nurses have always worked with limited resources, always had to practice within unwieldy institutions, always had to balance the patient in one room against those in the others. Those problems never got the attention of bioethicists. One person (not a nurse) with whom I talked said vehemently: "Put this in your book: Bioethics apes power! It follows doctors, not nurses." That is a significant reason these questions are not prominent on our agenda.

With a Philosopher's Tools

One more factor sometimes focuses bioethics' attention: the limitations on what philosophical skills can do. Not all bioethicists are philosophers, but many are, and some aspects of the discipline have been influential. Modern philosophical ethics, to the extent that it directly engages action, tends to do so at two levels: individual actions, or the shape of society as a whole. For centuries it was blind to intermediate structures amilies, communities, institutions nd so has few tools for understanding and evaluating them.

I don't want to paint too simple a picture. Despite the way economic, institutional, and disciplinary forces converge, other factors sometimes counter them. Recently patient-rights groups (especially AIDS activists) forced attention to the right to be included in clinical trials, not just the right to abstain. Currently serious attention is being directed to the conflicts of interest that arise when biotechnology invests in scientific research. It's probably also true that the appeal of Doctors' Dazzling Dilemmas is waning, as fresher issues summon bioethics attention. But Science Sensations will always be with us.

Some Topics We Ignore

One of my interviewees believed that "we drift like a tide with whatever is hot." She illustrated her point by noting that national health care was virtually abandoned as a topic after the Clinton proposal failed. That is one of many examples of inquiries not launched, or abandoned too soon. To appreciate the sometimes peculiar boundaries of bioethics, it helps to think about what is outside them.

Uninsured and the Poorly Insured

Roughly 45 million Americans have no health insurance neither private insurance nor Medicaid nor Medicare. Because Medicaid reimburses poorly, the additional tens of millions it covers often have great diffculty finding a doctor or hospital who will treat them. In one town in Michigan, no obstetrician will accept Medicaid patients; women in labor must drive fifty miles to the doctor they've been seeing, and a country hospital en route describes itself as the place where women stop when they can't make the whole distance. In big cities the concentration of the uninsured and the poorly insured can make the hospitals who treat them insolvent; the hospitals close, the burden is shifted, other hospitals farther away take it up for a while but for the sick poor, "farther away" can be unknowable and unreachable.

Bioethics has consistently paid some attention to the uninsured. Textbooks for undergraduate bioethics courses include units on "justice," focused on the question of whether there should be national health care or insurance. Philosophy has an extensive literature on questions of distributive justice and most philosopher ethicists, although not all, concluded long ago that health care should be a right rather than a consumer good. But once that conclusion has been reached and here is the point philosophy has little more to say. The subject does not usually come up in medical school courses, nor in workshops or presentations to health-care audiences. The poorly insured get even less attention.

Things were different for a brief period of time in the early 1990s, when health care became a significant issue in electoral politics and Bill Clinton proposed a form of national health care. Bioethicists at that time said and wrote a great deal about the subject, and some took part in the task forces that drew up the Clinton plan. Afterward, those I know dropped the subject for years. Partly this was because audience attention had gone elsewhere; it was about that time that managed care became such a painful issue and the human genome project was breaking ground. But I believe that a sense of hopelessness, even heartbreak, also contributed to the silence. Some of my closest colleagues had worked a lifetime toward universal health care, and had seen the best chance of a generation fail.

The plight of those with little or no health insurance, then, fails the tests of what is likely to get bioethics attention: it is not a problem faced by most doctors in their daily work; there is nothing new or exciting about it; there is no public debate creating an audience for discussions of it; philosophical ethics sheds little light on its cause or on the practical pursuit of a cure. I would have hoped that at least the question of charity care would arise do doctors and nurses have a moral obligation to spend part of their time giving uncompensated care? I broach the question, at least, in my work with third-year students, and offer material on it to the faculty teaching them on our community campuses around the state. Neither students nor faculty express interest.

Social Determinants of Health

Still less frequently mentioned is another fact of great philosophical and moral interest, but so little subject to individual action or even to legislation that it falls outside every bioethics category. That is the profound influence of socioeconomic status on health. The influence goes far beyond the obvious fact that poverty correlates with poor health. Audiences are never surprised to hear this, and generally attribute it to lack of health care and to bad health habits. (The first is often true; the second may or may not be.) Two other major contributors are virtually ignored, I believe because the American individualist lens is so powerful. That is, we tend to explain everything in terms of characteristics of individuals, and to be blind to the fact that systems have properties that individuals within them do not.

The first overlooked factor in the health status of the poor is the nature of poverty itself, which makes it hard to deal with even the ordinary details of daily life. Money is correlated with, indeed essentially *is,* the ability to control the circumstances of one's life. The poor must worry every day about whether they can keep warm, pay the rent, feed their children, and get to work. Not much on this list is reliably within their control, and a sense of control has frequently been correlated with health status.

What we know about heart disease illustrates this point well. Feelings of hopelessness, despair, and failure correlate strongly with the risk of death from heart disease. Some studies show that these factors are considerably more significant than better-known risk factors like raised cholesterol levels and hypertension. In the American context these conclusions are all too likely to be translated into facts about individual psychology. The terms "despair" and "depression" are easily interchanged, and when that is done, certain mantras arise automatically ("depression is a treatable disease") and certain causal explanations begin to be assumed: it's a chemical imbalance in the brain; it's a result of your genes; it's the way you were raised.

But it might also be a result of your social situation. Some people really do have less control over their lives, and less to hope for. One study linking hopelessness with the risk of heart disease found "depressed affect and hopelessness more common among women, blacks, and persons who were less educated, unmarried, smokers, or physically inactive." Note the way the sentence structure suggests that these factors are all the same kind of thing, all properties of individuals rather than of relationships among individuals. But gender, race, and marital status are essentially relational, as is "*less* educated."

Even emotions (like hopelessness and despair) are intrinsically more relational than we usually recognize. As one analyst notes, emotions "are more than psychological states. They are also reflections of culture and social structure. This is why, for example, anorexia nervosa, a form of self-imposed starvation [accompanied by certain emotions and beliefs], is found almost exclusively among teenage girls and young adult women in western industrialized societies, while *taijin kyofusho,* a morbid dread of causing embarrassment to others, is observed primarily in Japanese society. Feelings of despondency, depression, and despair are now understood to be so culture specific that guidelines for psychiatric evaluation recommend that cultural background be considered in diagnosing them." Similarly, culture will determine what counts as success and failure, and powerfully shape what beliefs about the future count as hopeful or hopeless.

There is much more to understand about the health status of the poor than whether they have access to doctors and whether they eat wisely and exercise. One way to reinforce this point is to note that in the United Kingdom, where health care is universally available, and where health statistics are collected by socioeconomic class, there has been a constant correlation between class and health. It's not just that the poor do badly. Those at each of the five steps do better than those one step below them. It's a steady, staircase descent.

Finally, there is a great deal of recent evidence showing that the degree of inequality within a society *in itself* contributes to poor health, and equality to good health. Of two societies with the same median standard of living, that with less difference between rich and poor will have the better average health. The affluent in a society with extreme inequalities will be less healthy than the affluent in a more egalitarian society as will the middle classes and the poor compared to their counterparts. This leads to surprising, even ironic, conclusions: for instance, the best way to improve the health of the rich may be to lessen the degree of poverty. And the correlation helps to explain why Costa Rican health statistics equal or outstrip those of the United States.

As interesting as these facts are, one may wonder whether they have any relevance to bioethics. After all, some stratification is inescapable in societies of any size, and the extreme stratification within the United States is not the result of anything doctors, nurses, or patients do. My claim, however, is that part of our job is to improve the moral perception of our various audiences. If Americans tend to be blind to one of the most morally important characteristics of our society, and a characteristic that sharply impinges on health, then bioethics should at the very least be helping us see.

Hospital Workers

I have already mentioned the fact that nurses can find themselves morally compromised. The same is true of dietitians, social workers, and anyone whose work with patients is subject to the authority of others. When the problem arises about the care of an individual patient (say, tube feeding for a patient who seems neither to want it nor to benefit from it), mainstream bioethics offers some help. An ethics consultation can be called, and there is a significant body of literature addressing the question.

For other situations, ethics consultations offer no help. Think of a chaplain, told by a hospital vice president to make a "home visit" to a recently discharged patient, a state legislator. The chaplain believes this is unjust, that his limited time should be spent helping ordinary patients still in the hospital. But he has little choice in the matter. Or think of nurses on units where tasks once considered part of nursing have been reassigned to housekeepers and orderlies. Such "restructuring" was common in the late 1990s, since nursing salaries are a large part of any hospital's budget, and since some traditional nursing tasks look unskilled: changing catheter bags, recording food consumption, moving patients, and so on. These reasons, in addition to the severe shortage of nurses now looming, lead hospitals to staffing patterns "restructuring," understaffing, mandatory overtime for exhausted personnel that nurses believe shortchange and endanger patients. One nurse in such circumstances said, "Every day I leave [the hospital] feeling that my heart has been torn out."

Such nurses suffer from what is called moral distress, a different phenomenon from the dilemmas to which bioethics typically responds. Dilemmas create puzzlement; they arise when it is not clear what is the right thing to do. In contrast, moral distress is a sense of complicity; it is not uncertainty about what is right but the belief that one is virtually forced to take part in what is wrong. Moral distress is likely whenever one human being is subject to the authority of others; it is not a symptom of evil but a sign of the human condition. We are social beings and we are moral agents: because we are social beings, most of what we accomplish depends on the coordination that institutions (with their hierarchies) make possible; on the other hand, because we are moral agents, we can never simply agree to follow orders. Bioethics offers little on this topic. A few pioneering writers in the 1980s named the phenomenon, but little further work has been done; and the field offers nothing in the way of practical help.

These are not the only interesting ethical questions concerning hospital employees. They can be exploited, for instance; and conversely, they can use their own power unfairly. Should lab technicians strike? Should employers try to block or break unions? Should nurses bargain for twelve-hour shifts, knowing that these schedules are not ideal for patient care? (Most nurses are women; women are likely to do most of their families' child care and housework. Because of this "double day," borne by women around the world, American nurses sometimes press for, say, three 12-hour shifts a week rather than five or eight-hour ones.) When exhaustion sets in toward the end of the day, patient care may be compromised; furthermore, scheduling diffculties become formidable. When some are on twelve-hour shifts, some on eight, some possibly on four, some working three days a week, or four, or five under these conditions, real teams, groups who know one another and work together smoothly and efficiently, cannot emerge. If bioethics has little to say about how hospitals should treat workers, it has still less to say about how the workers should treat the institution.

Nor do we have much to say about how health care professionals should treat one another. One nurse manager described a pledge he periodically asked his staff to take: "Some people find it hokey, especially 'I will not engage in the "3Bs" [Bickering, Backbiting and Bitching].' And it's not a magic bullet. It just puts into words basic decent human behavior. But how often do we do that? Once in a while I'll suggest, in general, the need for forgiveness; but that's a higher-level ability, not everyone can manage it, and I never ask it of individuals."

He has a moral goal, to help his staff work together in mutual respect; and he understands the difference between basic decency and higher ideals. He's engaged, in other words, in ethics work. The tools he uses emphasize talking with those with whom one has trouble. When I talked with him he had just concluded the yearly performance reviews for his floor, in which he discussed with everyone, and had each once again sign, a "Commitment to My Coworkers" statement. "I've felt a real lightening of our spirits since then; energy has been freed up that can go to patients. That's the way I sell this to my staff." His unit also had remarkably low turnover. Low turnover means experienced staff, experience brings with it effciency, and the result is that nurses have more time to review the plan of care and more confidence in approaching the medical staff with their concerns.

Patients benefit when there is mutual respect among the staff. But this nurse manager knew that mutual respect is also a moral good in itself. His help in understanding and fostering it came from management literature, not from bioethics. The electronic search engine Bioethics line in June 2000 listed over seven thousand articles on confidentiality, only two of which were about gossip.

Patients' Perspective

One young physician gave me his personal "genealogy" of bioethics: "First there were the precursors, like Joseph Fletcher [theologian author of *Situation Ethics*] and Henry Beecher [physician who blew the whistle on medical research abuses]. Then there was the first generation, people like Dan Callahan [founder of the Hastings Center], and next the generation trained by those pioneers.

It seemed clear to him that the field, at least in its clinical aspects, had finally reached its maturity. The proper people were at last in charge. Nonclinicians had been indispensable as pioneers, but now the torch had passed. (If pressed, I believe he would have included nurses, therapists, and others from the bedside as acceptable in a mature clinical bioethics.)

A nurse from the same institution agreed with this description, but not with his endorsement of it. Her remark, to the contrary, was that "the patients' perspective is dropping out of bioethics." If, as I argued earlier, bioethics tends to see issues through the eyes of physicians, her remark is not surprising. Note that she didn't say patient *issues* are dropping out of bioethics, but that their perspective was. She went on with an example: "Most of the debate about assisted suicide is about what it's proper for a doctor to do. It's not about what patients need and want. Or deserve." Once again, my own files attest to that: they contain a folder called "PhysicianAssisted Suicide" and carry articles about Jack Kevorkian, about the Oregon law and the Supreme Court's response, and articles by ethicists commenting on all this. I have no folder labeled simply "suicide" or even "assisted suicide."

The perspectives of doctors, nurses, and therapists are essential in bioethics. They understand things no one else can, and can speak to one another with force and candor. But their perspectives are not enough; some important concerns are only visible from other angles. Take, for example, the question of assisted suicide. Margaret Battin's "Assisted Suicide: Can We Learn from Germany?" describes a cultural and legal context very different from that in the United States. American attention has generally gone to Holland and not to Germany; in Holland, physicians may and sometimes do perform euthanasia. Germany deals with the issue quite differently: as a result of its Nazi past, it rejects euthanasia out of hand, but suicide is not illegal, nor is assisting it. It is not doctors, however, who provide the assistance. Suicide among the very ill does happen, not uncommonly, but it is almost always a private action rather than a medical one. There is a national society dedicated to helping those who want to end their lives, something like the Hemlock Society here.

Battin's article suggests a number of moral and ethical questions, including ways of understanding suicide and the proper role of private organizations formed around the issue. She is careful to talk about the subtleties and difficulties of trying to learn from other countries. She frames the article, however, within the pressing practical legal question of whether physicians in the United States should be legally empowered to assist with suicide. This is natural in many ways: the legal question being debated was about doctors, not about private organizations, and bioethics has always been driven by practical and especially by legal questions. In another way, however, her emphasis is odd. Because the configuration of law and custom within German society is almost impossible to imagine here, the more useful focus of her discussion would have been the proper roles of "aid in dying" societies. About this issue little has been written. Nor is there much about the moral nature of suicide, its justification or phenomenology. The question so heatedly and thoroughly addressed is "What should doctors do?" This discussion offers no help to the patient who might be trying to think through her situation, nor to the layperson to whom she might turn for help.

Another example is that physicians can be oblivious to the fact of, and the moral relevance of, health-care disorganization. I am not speaking here of insurance companies and managed care, both the target of constant attack. Instead, I am talking about the kind of problem that threatens any complex organization. The movie *The Doctor* illustrates this beautifully. The central character, a physician with cancer, is shuttled from one office to another, one clinic to another, until he finally erupts: "Is there a rule in this hospital that everyone who deals with a patient has to begin by saying, 'I'm sorry, but' you'll have to give us this information again, on this other form. I'm sorry, but your records have been lost. I'm sorry, but there was a misunderstanding about the time of your appointment."' So many doctors are involved in the character's case (and again, this is typical) that no one is quite sure what he has been told. As a result, he learns bad news inadvertently: "As you know, your cancer is growing." One physician ethicist with whom I work in Lansing frequently remarks, "What this patient needs most is a doctor" meaning someone who will unify the fragmented care the patient has been receiving from rotating teams and crowds of consultants. (In a teaching hospital students and young doctors move to new units every few weeks.)

One of the most basic concepts in bioethics, that of suffering, should tell us that these facts have moral relevance. Suffering is not the same as pain and injury; instead, it is a way of experiencing them, which follows closely from the meaning one attaches to them. Eric J. Cassell construes suffering as the sense that one's identity is threatened, that disintegration looms, often beginning with the inability to accomplish those things one considers important. Cassell notes that "even in the best settings and with the best physicians, it is not uncommon for suffering to occur not only during the course of a disease but also as a result of its treatment." I do not know that he had systems failures in mind; but, in fact, for anyone confronting illness, being able to talk with a doctor, get tests done, find out the results, perhaps choose treatment but certainly have a say in how it is scheduled, are highly important. To be systematically frustrated in them is a kind of suffering, but one that doctors have trouble recognizing. "To be effectively treated," Cassell remarks, "suffering must be recognized, but that is not a simple matter." In fact, as I discuss in the chapter on moral development, it is a moral accomplishment. Research suggests (and any conversation with laypeople will confirm) that dysfunctional systems are common in health care. Doctors are not always aware of these problems, and do not always take them seriously. To some extent this is the result of their training, focused on the deep technical intricacies of biological functioning and on the quality of their personal, individual interactions with patients.

Systems failures, then, and clinicians' overlooking of the failures, are areas where the patient's perspective is important, but too often missing, in bioethics. There is a crucial, unbridgeable difference between what clinicians witness and what patient and family experience. Another example of the way patients' perspectives can be erased is one I've mentioned earlier, situations described as "maternal-fetal conflict." The label expresses the clinician's perspective and erases that of the pregnant woman, for whom the conflict is between herself and the health-care professional.

Other Examples

Many other ethically important topics receive scant bioethics attention. Some are mundane, some not: Under what conditions should one agree to have someone involuntarily committed, given that conditions besides psychiatric illness can underlie disorientation, and that facilities vary widely in quality? What is the most responsible way to reduce and dispose of medical waste? How do reimbursements to oncologists influence their recommendations about chemotherapy? One young oncologist believed the influence is corrupting: "Watch the language with which the chemotherapy is offered. Patients will be told, 'We don't know for sure that this won't work for you.' What they should be told is, 'There is no reason to believe this will help you, and a great deal of evidence that it won't.'"

Interestingly, the questions raised by this young oncologist fit the pattern of classic bioethics issues; that is, they concern decisions made by doctors about the care of individual patients. Perhaps these questions are not taken up because not a lot of doctors find them distressing, and nothing has happened in the public arena to make them seem problematic. Their invisibility speaks to other kinds of questions the silencing of nurses and young doctors, the ways in which the media accept and perpetuate a certain stance toward cancer.

Practical Steps toward Choosing Well

The virtue of choosing projects well is required if we are to help people see such issues. It requires keeping in mind the field's moral occupational hazard: whenever we choose to focus attention on one issue, we in effect draw it away from others that may be more important. Putting that concern aside, however, even work properly focused on a certain question suffers if the fabric of related questions is not recognized. Bioethics is like a set of stakes pinning down a great billowing canvas. Some areas are so thoroughly anchored that the images on them can be read, evaluated, and corrected if necessary. Other parts of the canvas surge and undulate out of reach, threatening to loosen what has been anchored, and making even that hard to read. Context is important, and some of the habits of bioethics obscure it.

Choosing projects well demands courage, discretion, and something like night vision, the ability to see when the light is dim. Although there is no formula for developing it, there are some rules of thumb. The factors described above provide a useful set of questions with which to begin: we can develop the habit of noting whether we are too bound by the problems of physicians, the dazzle of dilemmas, the lure of funding, and so on.

Listening to the "Other"

Beyond asking ourselves those questions, however, it is also useful to seek out the voices of outsiders and of the young. Thomas Kuhn suggested long ago that those not fully socialized into a field have the most original vision. The story of our medical school curriculum, whose initial focus on death I discussed earlier, is an example of the usefulness of outside perspectives. A few years ago we added to the curriculum some minority-focused issues, including the Tuskegee syphilis study; that quite naturally led to material on clinical trials in general (in the form of questions faced by primary care physicians). This inclusion was a first step away from our too death-focused curriculum. Attention to those who tend to be excluded, in other words, should not just be seen as an act of justice; it will often broaden our understanding of apparently unrelated issues. Another instance of this broadening occurred within the Joint Commission for Accreditation of Healthcare Organizations (JCAHO), which has enormous power over what hospitals do. Until about thirty years ago, its standards emphasized patient welfare but said nothing about patient rights. (The difference is important when a patient wants something that hospitals believe is not in his or her best interests.) The standards began to change in 1969 when the Joint Commission sent out a revision of its standards to a number of different groups for public comment. This was simply a move toward hearing from a broader set of people. And it was the National Welfare Rights Organization (NWRO) not doctors, nurses, or hospital administrators who proposed including patient rights. NWRO was an umbrella organization comprising more than 300 local welfare rights groups, whose members in turn were welfare recipients and welfare activists. Once again, listening to outsiders and minorities led to a broader understanding of one's responsibilities generally.

The National Bioethics Advisory Commission (NBAC) did something similar. Its mandate was to "provide advice and make recommendations" about the "appropriateness" of governmentally sponsored bioethics research; to "identify broad principles" for the conduct of such research. It would be easy to interpret that mandate as calling for the same kind of work one would publish in a bioethics journal, focusing closely on what makes cloning, for instance, morally permissible or impermissible. But NBAC wanted to hear a variety of perspectives, and invited a variety of speakers, including those with theological commitments. This move received a certain amount of criticism: religion is a private matter, maintain the wall of separation between church and state, and so on. But the result of listening to such voices was a significant reconstruction of their job: not to make recommendations about the rightness and wrongness of certain practices, but to make recommendations about *ethically defensible public policy in a deeply divided democracy.* That means paying attention not just to, say, the legitimate use of stem cells, but also to what it means to respect a minority. It means asking about more than the moral status of the zygote; it means asking about the moral distress of those who oppose the proposal, refusing to dissolve that distress into mere preferences; being suspicious if the benefits from a program accrue only to those who approve of it; asking whether objectors would be forced to become complicit, asking whether they would be allowed to use democratic process. Once again, listening to the "other" broadened the moral landscape in unexpected ways.

Even when the conversation is confined to "insiders" to those explicitly involved in health care, or to those in the majority it is important to hear from a variety of professions and types of institutions. For one thing, each highlights a different aspect of a problem (biochemistry, social situation, the impact on a sense of meaning); in addition, each includes a set of conventions, of habits and norms, that can be set in illuminating contrast with one another. I began to see confidentiality in a new way, for instance, after hearing nurses and social workers talk together about it. The two fields had quite different customs about how much patient information to share *with one another.* Before, I had thought of confidentiality as it is usually presented in bioethics, as a question of how much patient information to reveal to laypersons, especially to family members and government agencies. Now I began to think about the way information flows within a hospital, and to see that the currents of talk (about patients and about one another) are virtually unimpeded. This awareness brought to mind comments I had heard in other contexts. During a discussion of testing hospital employees for HIV, for instance, someone in the audience had said, "If one of us were found HIV positive at our lab, the news would be all over the hospital within twenty-four hours." Assent swept through the crowd. Having heard that public health departments protect patient confidentiality better than hospitals do, I talked with nurses from that setting about the question. They agreed with the impression I had, gave me examples of the lengths they go to protect patient confidentiality, and said that the stricter norms apply to employee information as well: "We don't worry about information [even about one another] leaving the lab. It won't." I use that anecdote in speaking to doctors and nurses, to show that different customs are at least possible.

For the same reasons, interdisciplinary work is crucial. Each discipline sheds light on a different aspect of life, has its own rigorous tools and its own skepticism. The sociologist sees patterns of power and the detailed structures within institutions, not just the political and economic factors of society as a whole. The anthropologist is able to find the meaning that situations carry for those involved. In a later chapter I will talk in detail about the importance and the challenge of good interdisciplinary work.

Finally, attending to a variety of religious perspectives is enlightening. Hospitals founded by Catholic religious orders, for instance, often had departments called something like "Mission Effectiveness." These departments had a voice in a wide variety of institutional decisions, including mergers and acquisitions as well as care for the poor a breadth of concern far broader than hospital ethics committees are accustomed to taking up.

Seeking Long and Wide Views

Another useful tack is to take a long view, to track the way a question has appeared over the years. More than one person, for instance, believes that within ethics consultations a pendulum has swung, that, whereas ten years ago the common problem was patients or families wanting less intervention than doctors did, today patients and families tend to want more. In another instance of this kind of shift, I was startled recently to hear medical students talking about women who wanted cesarean sections their doctors did not believe were "medically indicated." The issue used to be just the opposite: women refusing c-sections their doctors felt were imperative. Both shifts (or apparent shifts; I have only anecdotal evidence of either) led me to wonder whether patient preferences trail medical opinion by about ten years, a certain kind of socialization of the public taking place largely through the media so that patients finally come to want what doctors think they should want, just as doctors move on to a new standard of care. The whole picture (of what gives rise to disagreement between doctors and patients) is obviously more complicated. But noticing this curve has made me still more attentive to the power of the media, and more concerned about the ways in which "health promotion" campaigns can go wrong. I would not discuss doctor-patient disagreements without taking this apparent historical trajectory into account. Obviously the deeper and more critical view of professional historians is still more important.

It is also useful to try to see any specific issue as an instance of a broader one. To continue with the example of confidentiality, once I began to see the standard issue (about revealing patient information) as only one of many questions about the way information flows within health care, information not just about patients but also about workers, I came to realize that while the flow is too free within hospital walls, arguably too little information leaves those walls. It may be common knowledge within a medical community that Dr. Y is incompetent, unethical, or even criminal, but patients will be the last to know. The question of who should be told is hard, of course, but important, and like everything else I've mentioned, it gets very little bioethics attention.

Being an Outsider

Finally, a number of thoughtful commentators have mentioned the importance of remaining in some sense an outsider oneself. The sociological literature cited here contains a number of different terms and models, as does the less frequently cited but still more interesting feminist literature. The early sociologist Georg Simmel remarks that being a stranger entails a certain degree of contact: inhabitants of a distant planet are not strangers unless we meet them. His archetypal stranger is a trader, whose firmest roots are elsewhere but whose interaction in the territory is systematic and ongoing. This trader is freer to leave than the natives with whom he interacts, is outside many of the bounds of convention, and has "objectivity": "a peculiar composition of nearness and remoteness, concern and indifference." The stranger is often chosen as a confidant (a point that triggered for me memories of listening to an anguished neonatologist). Writing somewhat later, sociologist Alfred Schuetz uses a different paradigm for "stranger," that of an immigrant, or any "applicant for membership" in a closed group such as a family, club, or nation. Where Simmel spoke of objectivity, Schuetz speaks of disinterest: the stranger "intentionally refrains from participating in the network of plans, means and end relations, motives and chances, hopes and fears, which the actor within the social world uses for interpreting his experiences of it. As a scientist he tries to observe, describe, and classify. The actor within the social world, however, experiences it primarily as a field of his actual and possible acts and only secondarily as an object of his thinking." Edward A. Tiryakian, still later, remarks, "'Stranger' is not a static category but one that any one or any social group may move into and out of, most typically by wandering or migrating. . . . But colonization also produces a different kind of strangers, those who have become strangers in their own land." All this supplies a rich lode of material for thinking about being a stranger, an "outsider within." This last insight is crucial to feminist writers like Patricia Hill Collins, who coined the phrase to describe the situation of African Americans and who points to the epistemological advantages of the situation. Gloria Anzaldua and Maria Lugones employ powerful metaphors to convey the ways in which one's very identity can be mixed, troublesome, enlightening, and useful. (Anzaldua: "This Bridge Called My Back"; Lugones: "curdled identity.")

These approaches cannot be homogenized. Obviously there are many different kinds and degrees of outsiderness. It was important to Simmel's first "stranger," the trader, that he could leave; his primary identity was elsewhere. The feminist writers are talking about quite different situations, situations where leaving is impossible. Yet there is a common thread among the theorists: distance and difference can help one see, but do not necessarily help one be effective. Outsiderness translates into power only when it provides some independence and when insiders *value the perspective* it provides. The young oncologist with whom I talked, who had turned from curative to palliative care and to ethics, was mocked by his former colleagues: "You really like holding patients' hands?" Nurses who learned to speak up, in one study, *never* found that doing so made a difference in patients' care.

Nevertheless it is true that the right kind and degree of outsiderness can be useful. I talked with one consulting bioethicist who was a "stranger" in several senses: He was not a clinician, and not a regular hospital employee, since he was contracted out by his home institution and had other duties elsewhere. The hospital board of directors had approved his hiring and charged him with "being visionary." His official job title was simply "ethicist." That combination of visibility and "outsiderness" meant that people came up to him with a wide variety of concerns whatever struck them as having to do with ethics. He offered this example: "We include a brochure about ethics in the literature we give patients; it's very general, leaves the word undefined. One day someone came up to me to complain about the food not exactly an ethics question. I thought. Then I realized it was really a concern about fidelity. He had been promised something, face to face, by a hospital employee."

This man was an outsider in a number of senses: he had the independence of Simmel's trader, as well as the legitimacy. Furthermore, his language was to some degree a pidgin: listeners thought they understood the word "ethicist" but interpreted it more broadly than most "ethicists" do. All this made it possible for him to see and to take up unconventional questions, even those that could be threatening to those with power. He defined ethics much more broadly than has been usual, in part because of the religious tradition from which he drew: he saw employee relations, for instance, as an area in which abuse of power was possible. In addition to recognizing that fact, he also took action, drawing up a questionnaire assessing the "moral climate" in which employees worked.

Choosing projects well, then, if it is not a virtue, is at least a moral excellence, one crucial to our work. It demands understanding the forces that influence the field and the ability to respond intelligently sometimes resisting them, other times using them as sailors do the wind, to move in a different direction. The virtue demands seeing what is overlooked, a vision that is helped by listening to a variety of voices and by being, in useful ways, an outsider within. But it is not just a matter of seeing; it also involves doing. As one of my interviewees remarked, it calls for a most unscintillating virtue: time management. He gave a mundane example, but one that shows he had thought through his priorities. "I get more speaking invitations than I can accept. I'll choose professional audiences over, say, church groups and the Elks, the kind of groups that have a speaker every month on 'Whatever.'"

11

Bioethics as a Territory in Life Science

Introduction

People taught and wrote about ethics for doctors, for nurses, for therapists, for scientists. Their domain, however, was significantly different from what we now think of as bioethics. In most instances, these people were clinicians or scientists themselves, articulating norms into which they had been socialized, and exhorting their fellow professionals to follow them. Some came from religious traditions; they, too, were primarily repeating and explicating a set of norms shared by their readers, norms that changed slowly if at all.

Discovery

Suddenly this land attracts interest from outsiders. Some dramatic discovery draws attention to its possibilities: perhaps gold is found, and passages are forged over mountains here- tofore impassible. Those who happen to be in the old camps when the newcomers arrive react variously; some welcome the interlopers, others try to fight them off. It becomes increasingly clear that the land holds more promise than has been noticed before, and that the ancient kingdoms might profit from its development.

The gold in this case was a set of new questions that offered intellectual interest, a chance to make a difference in the world, and the possibility of professional advancement. And for the first time these questions seemed to belong to the public domain. Sometimes professional barriers were lowered voluntarily and the newcomers welcomed, but at other times the walls had to be stormed. Medical scientists had not been adequately protecting the subjects of their research; doctors and hospitals did not want to decide by themselves who received scarce and life-saving kidney dialysis; in an era of civil and consumer rights, patients were growing dissatisfied with medical paternalism.

Immigration

A trickle, then a stream, of newcomers arrives, from many different homelands and for many different reasons. Some are missionaries, some homesteaders, a few are tourists, and a few are hustlers. Some are wandering aimlessly and stumble in. There are many hungry and adventurous young, and a few elders, hesitant, or thoughtful, or overbearing. Many have not been quite at home in their countries of origin.

The people who began to write, teach, speak and otherwise do the work of bioethics were not professionally trained in it. They came from religion, medicine, philosophy, law, nursing, the social sciences, and literature. Some came with a moral mission, a response to scandal or to some aspect of their own profession that disgusted them: "In residency once I was holding the bag the CPR bag for a demented woman in terminal pain. Nobody had a clue whether she would have wanted this or not. I swore I'd never be in that position again." Among those who had dealt with illness in their families, the quest could be moral in the broader sense, a desire for a deeper understanding of what had happened. "I was appalled at the way the desire for money influences the profession."

Some stumbled into the territory never having sought it out. Many of the people to whom I talked described the chance events that brought them to this new arena: They knew people who were taking on projects, or came across a grant opportunity, or "put themselves on the market" and ended up "sold" to medical ethics.

Someone started the center I got into it accidentally. In graduate school I impressed my adviser as a good writer, so he hired me as a grant writer in the center. Some were professionally homeless. People who had spent five or ten years of their lives earning a doctorate emerged into a new academic world, a fiercely competitive arena in which the losers were not simply undistinguished but unemployed. The humanities had entered an economic depression from which, thirty years later, no recovery is yet in sight: Many more people earned Ph.D.'s than could find positions in the field; those who did were given teaching loads that left no time for laundry, let alone for research and writing. By 1970 the departments who hired them were becoming hungry for enrolments; in the 1960s people had been willing to major in liberal arts, but as the Vietnam War wound down and the economy slowed, students turned to majors they thought would help them earn a living. The fortunate were hungry in a more traditional sense: young scholars who needed to publish, and hoped to build their careers around some new and exciting area of inquiry.

Others had a place to stay, but wouldn't call it home.

I taught high school one year. After the first day I cried all night. I cried a lot that whole first year. I had to find something else.

I hated [that university]. Yes, I was lucky to have a liberal arts job, but such conservative students and the community, too; I was literally spat upon for protesting Vietnam. The students were not interested in what I did.

That speaker gave up a tenured position in order to move into bioethics. Another gave up her chance at tenure: "In philosophy you're expected to be grateful just to have a job, any job, anywhere." But she, far from being grateful, was actively unhappy. Someone else, from another of the academic humanities, felt not quite at home there intellectually or politically: "I was more interested in meta issues. I like to work with cognitive dissonance. And like the people I work with so many 'old lefties' in the crowd we want to retain some sense that we haven't entirely sold out."

Some were prospectors: academics, for instance, struck by the possibility of new questions and forms of work life.

In my late 40s I took stock. What resonated with me was activity that makes a difference. I continue to be dismayed by the narrowness of my philosophical colleagues.

I'm a practical person. [I'm happiest doing something useful.]

A lot of people go into philosophy in order to ignore the world. In philosophy of physics we were allowed to look at the real world.

But then I found out that philosophers of science didn't really know how science was done. I liked logic, but didn't like doing philosophy of physics as if it *were* logic. I couldn't get interested in Judith Jarvis Thomson's way of doing ethics through wildly fanciful analogies.

Some who came from medicine were moved by intellectual hunger.

I was looking for some place to think. An antidote to the usual routine in medicine which is "Do, do, do and think. Maybe."

I was numb from years of school and residency, and then suddenly I was out of the clinic, with time to do whatever I wanted.

I began working in a facility for the chronically ill, ventilatordependent, often somewhat demented; the families could be demanding. All the [other] doctors and nurses hated the work.

These patients were the cast-off "successes" of our medical system. *What was this all about?* Suddenly I realized medicine could be interesting again.

Other physicians needed not stimulation but resolution. "I would have left medicine if I hadn't discovered bioethics. The medical part is easy: You look it up, and after seven years you should know how [to do almost everything]. But when I encountered what I came to call moral issues, I'd get a different answer from every senior physician. I needed some way to be sure that I had included all the significant factors when I made my decision."

Among the professionals who were young and came to the field when it was already partly developed, many report something like a conversion experience.

This was where my heart was.

I made my decision during a trip around the world. I wrote the letter in a tent under the stars, and I knew it was momentous; I made a copy for myself.

I just fell in love. I couldn't get enough.

Some were tourists. The mysterious and exotic world of medicine suddenly seemed open to select outsiders. A few have come to the field as hustlers. The field offers the opportunity to make a name or, if not quite a fortune, more money than in the usual academic life. (The contrast does not hold for medicine. A doctor mentioned that bioethics was one of the few specializations likely to reduce one's income.) Without credentials and standards, anyone could hang out a shingle.

Most, of course, arrived with mixed motives, somewhere on the spectrum between missionary and entrepreneur. They hoped to do well by doing good.

Settlement

Because the base camps are not adequate for the logging, mining, and farming that begins, the newcomers put up their own cabins; after a few years some of them make the new land their home. Some retain their original citizenship, but most maintain some kind of dual allegiance.

A few theologians and then philosophers took up the new questions. Some doctors and nurses did, too, in a different way than their forebears had done. These newcomers were not so much interested in transmitting accepted norms as in criticizing and changing them. At first these topics were only one of their many professional interests, and most retained their original disciplinary identity for a long time, unsure of whether this new "land" would sustain permanent settlement. Some, however, realized within a few years that they had developed a new professional identity, one that expanded or replaced earlier senses of self.

I'm always well grounded in anthropology; I haven't left it.

I don't write much for scholars any more; my audience is working professionals. My farewell to religious studies was gradual and easy.

I'd say I've abandoned social psychology; I haven't contributed any primary research for years.

First Years

Now that they form a group with similar interests, the newcomers band together for mutual support. Even the loners come into town occasionally for necessities and for company.

Settlements grow up around common projects: timbering, trading, farming, mining. At times there are temporary gatherings or the purpose of surveying the land, marking boundaries, and so on. Names for this new territory spring up in the way names do, pragmatically, accidentally, humorously. Some stick, but none seems quite right: the inhabitants are too different from one another, and the terrain too varied. The borderlands have less trouble, perhaps because they are smaller and more homogeneous.

People interested in questions of research and medical ethics talked with one another. Those teaching in undergraduate programs especially in philosophy, religion, or science departments developed courses for their own students. "Centers" and "programs" began to form within medical schools, with one or two core faculty and a number of part-time associates. Typically, but not always, these faculty developed courses for medical students. Task forces and commissions on the national level were less permanent, but often highly influential. Some centers formed independently of medical centers or universities, one or two people doing consultations about patient care, perhaps. Many different kinds of projects were undertaken: "white papers" for legislatures, conferences about the care of the disabled, changing professional codes of ethics. Many people began to sit on IRBs. Sometimes they did this work as volunteers; other times they made a significant amount of money for it. Hospital ethics committees grew up, and regional networks of ethics committees developed. Doctors and nurses developed pioneering ways of helping patients at the end of life, studied the results, and shared them with the professional and lay public.

The variety of activities is reflected in disagreement about an appropriate name: Bioethics? Medical ethics? Health care ethics? Medical humanities? Health and human values? My reason for choosing "bioethics" as the least inadequate is partly a result of being trained as a philosopher, where "ethics" refers to a broad field that encompasses all the moral dimensions of life.

Relationships

The settlements depend on one another. Some are near neigh- bors, others so distant they rarely interact. Some people own property in more than one settlement. Intermarriages take place. Trade and supply networks grow up. Personal feuds and territorial disputes flare; occasionally vigilante groups form. Some people become prominent, others are snubbed. Networks of privilege and prestige develop. Legends spring up.

A great variety in projects and in organizational structure characterized the field, hardly formed enough to be placed under one roof or name. Purposes ranged from the highly abstract to the thoroughly practical. People at one end of the spectrum did not always respect those at the other.

I don't especially want to schmooze with doctors; if I've got any extra time, I'll go into my offce, close the door, and think.

The moral work of medical ethics is local. Most of what is published is pretty useless.

An infrastructure developed: journals, professional organizations, discussion groups on the internet. In much of this groundwork, however, participants could sense the tension between theory and practice. The Society for Health and Human Values (SHHV), one of the first organizations, was considered too abstract by some who were daily involved in clinical ethics, and the Society for Bioethics Consultation (SBC) was founded to meet their needs. But SHHV was disdained by others for lacking theoretical rigor, and that reaction resulted in the formation of the American Association of Bioethics (AAB), largely by and for philosophers. By 1998 the three associations had merged. Some humanists, those in literature, art, and history, feared that the new consortium would push them to the margins, or entirely off the page, and considered starting a separate organization. The tensions took other forms as well; some social scientists felt philosophers were missing the whole point (and sometimes that philosophy, by its nature, could do nothing else but miss the point); some doctors felt the field should belong to them and them alone; some philosophers condescended to everyone (including other philosophers of the wrong degree of abstractness or practicality).

Feuds and pecking orders proliferated. To caricature the scene: Philosophers dismissed religion. Social scientists disdained everyone, including themselves. Legal scholars valued it all, including themselves. Nurses were invisible. Georgetown graduates believed they owned the field. The young knew they were the future. Pioneers knew that they'd forgotten more than newcomers would ever know. Philosophers assumed no one else could do conceptual work; social scientists asserted that observation is all. Bioethicists who worked in hospitals called their work clinical and considered themselves the elite, while, in a ranking of philosophy programs by specialty, not one clinical program was listed under "Best in Bioethics."

But sometimes, somehow, these groups managed to work together. Often the relationship was one of trade, but there was an increasing amount of cooperation on common projects. Some philosophers may have felt in the beginning that their mission was to enlighten others, but no one can do this work very long without realizing that they have as much to learn as to teach, and finally that everyone is learning together as they face a common problem. A neonatologist might call me to see what the ethics literature says about tube feeding; I might spend time with him to see the practical implications of a decision not to tube-feed an infant; we might work together to write a paper on the subject. As is true in many arenas of collaboration, even what looked like an exchange (of information, say) was more likely to be some form of gift, placing parties in ongoing relationships of mutual benefit. Just as a variety of motives brought people into the field, so a variety of human relationships grew up.

Borderlands

> *Borderlands begin to be settled, land that shares some of the attractions of the Territory itself, but adjacent to different Kingdoms. The Borderlands vary, but without exception are smaller and more lightly settled than the Territory itself. Borders separating these new territories from one another and from the old Kingdoms remain rough: "We have everything this side of the river, they have always governed the land beyond the grasslands but the river and the grasslands themselves? We both use them."*

While bioethics was growing up, so were similar fields: business ethics; many forms of professional ethics, especially engineering and legal ethics; environmental ethics and the animal rights movement; and so on. Some feminist theory also is of the same kind; that is, it aims at understanding practical ways in which the world could be better. Motives for entering these fields varied as they did in bioethics. One person, whose family, in her childhood, had been involved in arguably immoral activities, believes that she was deliberately shown some of the details and some of the results. "They hoped I would remember. I was a kind of messenger to the future." Her work now is an attempt to make public, and stop, what once flourished in secrecy. Another person's motivation is more generic. He reports, "I love a good fight: one for a good cause and with a real chance of winning." The "borderland" fields differ in many ways. While many speak to specific professional groups (to engineers or city planners, for instance), some take all of us as their audience. An example is environmental ethics: the minds it hopes to change belong to no single occupation. Many issues belong to no one field, or overlap many: with the advent of managed care, for instance, those doing "medical" ethics become interested in business ethics. Problems of confidentiality and conflicts of interest arise in all the professions.

Just as boundaries between these new fields are permeable, so are boundaries between them and traditional practices (the "ancient kingdoms"). For my purposes I will roughly stipulate some perimeters, bearing in mind that these are not locations of honor but merely of description: On the academic border, those who simply teach undergraduate courses (bioethics, literature and medicine, and so on) are outside, although just barely. Those who teach graduate or medical students, or who write for publication about the issues, are inside. On the clinical border, those who "simply" try to practice as ethically and humanely as possible are outside the boundaries, but just barely; those who develop what I call innovative practice are within. By "innovative practice" I mean programs, like a number of hospice and palliative-care programs, intended to pioneer new kinds of health-care practice. I count them as ethics projects because they are not just instrumental toward already understood goals, as new forms of chemotherapy would be. The goals of these innovative programs are themselves newly created; they have been identified, argued for, and refined. Furthermore, the goals are intrinsically ethical, since they have to do with what it means to respect persons.

The Need for New Skills

Newcomers find that they need new skills: for starters, how to clear land and what to plant on it. Some things simply won't grow; others grow at first but quickly deplete the soil, or die during harsher summers. And how does one cook without familiar ingredients? Or stay warm and dry in this climate so unlike the one at home? How do people find their way without roads? Conversely, old skills are of little use: highway driving, supermarket shopping, staying safe in a big city. The newcomers who arrived here as adolescents have less adjusting to do. But they cannot just follow the patterns their elders have set; there are no elders, in that sense. Together everyone must construct a new form of life.

The first years in bioethics were filled with trial and error. What should we try to do, and how? What efforts will bear fruit, and is the fruit worth the effort? Within a few years, ethics courses were deeply rooted in medical schools, but even after decades not within veterinary schools. Hospital ethics committees became increasingly important, as accreditation agencies virtually demanded them, but their effectiveness was increasingly questioned. Regional networks of hospital ethics committees were important for a while, but later many of them withered. National commissions and other intersections with government came and went. Electronic conversation lists were established. At first there was great excitement at this new way to communicate, but within a few years many of the original contributors dropped off, overwhelmed by the volume or unhappy with the dominance of a few garrulous contributors. In spite of those problems, the networks became effcient ways to exchange certain kinds of information rapidly, and good places for newcomers to learn.

Those trying to decide what projects to attempt were usually working within unfamiliar institutions: hospital ethics committees, regional ethics networks, medical conferences. Even medical schools were alien to many. For a range of new endeavors, new skills must be developed: from the relatively trivial (using slides for presentations), through the obviously challenging (managing a volunteer organization), to the deep (helping clinicians reach decisions about patient care; making recommendations about national policy).

Some of the needed skills were, and are, moral. Virtue, in fact, is a kind of skill: a habit of doing the right thing at the right time in the right way. Habits of honesty and kindness, of fairness, of maintaining integrity in the face of pressure and temptation one develops these through life in particular circumstances. Ways of being kind, for instance, differ as contexts differ; pressures on integrity vary as power structures vary. People in this new line of work occasionally find with dismay that they have compromised themselves, or fallen into dishonesty or unkindness, because the cues and challenges become harder to recognize. To take a single example, public speaking: What counts as manipulating an audience? When should one ask for a stipend, and how large should it be? Is there a name for the virtue of being able to look at audience feedback forms and dispassionately learn from them (rather than becoming defensive or distraught)? If one has little time, which invitations should be accepted, which declined? And how is virtue in these respects learned and taught?

MacIntyre characterizes virtues as qualities that are necessary for attaining the internal goods of a practice. (He also argues that courage, honesty, and justice are necessary for attaining the internal goods of any practice.) Aristotle thought of virtue as generally

"lying in the middle, " between excess and defect. Later I will look at the particular threats to integrity that this field presents, and explore the possibility that it requires particular virtues: discernment, humility, courage, honesty, and mutual respect.

The Young

The pioneers were not reared in this new form of life; they invented it. They carved space for it out of the wilderness. But, like primates who were not themselves nurtured as infants, some give no thought to supporting those who come after them. They cleared their own land; surely the newcomers should do the same? Others, however, form settlements that make everyone more productive and that easily integrate waves of new settlers as well as the next generation.

There is great disparity in the way newcomers to bioethics are treated. Some excellent things happen, both nationally and locally. The ASBH, for example, has an annual award for the best paper submitted by a graduate student, and gives the winner a steep discount in conference fees. In some graduate programs, students are proud and grateful for the help their distinguished teachers give them. Similarly, in some centers, veteran and new members meet regularly to discuss not only *what* they're working on but also *how, whether, and why* they're working together. On the other hand, there are programs where students are abandoned once they've graduated, if not before, and where junior colleagues receive only keys to their offce and good wishes. I take these issues up at greater length later. My conclusion will be that where there's a problem, it is in part a result of the newness of the field. Each center is *sui generis,* so bioethics lacks the kind of standard path that has long been laid out for medical residents, newly hired nurses, and junior faculty in established fields. Centers that assume new members should carve out their own jobs, as senior members did, can put the young in impossible situations, since all the obvious slots may already be filled. And the veterans may not notice that a newcomer is struggling, busy as they are and misled by an assumption they have never articulated ("I cleared my own land"). The young will, and indeed do, have a very hard time, waging an invisible and nameless struggle. Some will leave, and some will be forced out for "unsatisfactory performance."

Choosing Projects

After some years, the settlers become comfortable. In part they're self-sustaining – the climate is pleasant and certain crops grow well. Yet they depend for essential supplies on the older kingdoms, and they've learned how to keep those supplies coming. The need to please patrons, along with the habit of talking primarily with near neighbors, carries certain dangers. Will the community contribute to the world at large, or settle into cozy insularity?

Bioethics too became comfortable, and faced the temptations that come from insularity on the one hand, and from dependence on the other. The projects it takes up tend to be those that please the medical community, granting agencies, and general audiences. In later chapters I explore the moral challenges especially to integrity and discernment created by this need to please. There is also a problem of standards. New professional fora have developed associations, conferences, and journals but the work they generate can be lackluster. Journals of the "Medicine and X" variety can pale in comparison with the best work done in the original discipline. Since new journals address new questions, it is not surprising that they lack the depth and polish of the best work in older fields. But the contrast remains disquieting.

As years of effort accumulate, as opportunities to talk and learn from one another expand (a downtown grows up, in terms of my allegory), and if enough of us keep in touch with our disciplines of origin and with the outside world, all this is likely to change. In the meantime, I believe several things could help.

12

Bioethics as a Practice in Life Science

Introduction

A practice is a coherent and complex set of activities, socially constructed. It has distinctive goals and standards of excellence that help make the practice what it is, and that cannot be fully understood apart from it. Chess is a practice, checkers is not; architecture is, but bricklaying is not. A brilliant opening gambit in chess is a good example of a kind of excellence that cannot be understood if one does not understand the game; checkmate is a goal that cannot be understood without knowing what chess is about. These defining goals and standards develop through time; portrait painting, for instance, once aimed at symbolizing the subject (by means of objects arranged around him, say, and by his attire) rather than at capturing a good likeness; what counts as a good likeness also evolved, so that, beyond a recognizable image of the subject, what is wanted now is an expression of the inner person. Each practice is lived out in a specific way: the lives of a physicist, a rancher, and a portrait painter differ significantly from one another.

MacIntyre introduces the concept of a practice in order to talk about virtues. He argues that all practices, no matter what they are, require courage, justice, and honesty, but also that different forms of life, with their distinct practices, will differ somewhat in the virtues they require. Different eras of our own past, for instance, found different attributes to be morally praiseworthy. Homeric heroes needed different strengths than did Jane Austen's heroines. Benjamin Franklin praised thrift and hard work; Aristotle did not. Although these are clearly expressions of cultural difference, that fact does not end the moral analysis. Franklin's position *may* be simply ideology or moral propaganda, but it could also be argued that within capitalism thrift and industry are truly moral goods, because without them one is not contributing fairly to a system from which one will benefit. (The point would have to be limited, since capitalism needs criticism as well as support.) In any case, the distance between Franklin and the Greeks nicely illustrates the idea that different ways of life present different moral demands. My reason for examining bioethics as a practice is to look at its particular moral requirements.

Bioethics may not be suffciently developed to count as a fullfledged practice; perhaps it resembles bricklaying more than architecture. I will not tackle that question. For my purposes the field is close enough to being a practice to make MacIntyre's treatment illuminating. My major point is that, practice or near-practice, bioethics should be evaluated as a practical part of the world. It is not just a body of academic writing, and the practice of it cannot be evaluated adequately on solely academic grounds.

First, then, let me flesh out my sense that bioethics constitutes at least something *like* a practice. Its complexity is obvious. Our professional lives involve not just reading, thinking, and putting words on paper but many different kinds of learning and sharing. My own week in November 1994 was varied enough, but when we add what others do (and what I myself do at other times), the complexity becomes still clearer. Working on an animal-care committee, writing grant proposals, teaching medical students and undergraduates, writing for academic journals and informal newsletters, consulting on patient cases, presenting hospital Grand Rounds, helping federal agencies make policy, offering a listening ear to clinicians, organizing conferences, running networks of ethics committees, working with legislatures, modeling new ways of providing health care, speaking to the press, and so on and so on: this is a complicated life.

Its coherence is almost as obvious. The shape of each life in bioethics is somewhat different. Yet I think most of us find certain currents flowing through our activities: what we learn from practicing professionals, for instance, we share with students; challenges offered by students shape what we write. And so on, each activity informing the others. Later I will suggest some ways of conceptualizing these unifying currents, but now I will simply offer an example of the way the activities fit together by pointing to the issue called "physician-assisted suicide." Throughout the 1990s bioethicists responded to Jack Kevorkian, the Michigan physician who "helped patients die" and sought the limelight for what he was doing. The wave of deaths elicited a lot of bioethics activity: we taught, spoke, wrote for one another, developed *amicus* briefs for the Supreme Court, examined what was happening in the Netherlands (where euthanasia is accepted for certain situations), developed better forms of palliative care, and looked more closely at the way people die here. Legal argument, social science research, exemplary practice, philosophical exploration, comments to the press, workshops for clinicians: these are very different activities, but each was part of a coherent whole.

On the other hand, the coherence should not be exaggerated. Practices do not have Platonic essences, fixed and immutable. They are human enterprises, changing through time, shaped by the circumstances in which they grow. (Bioethical interest in "managed care" exploded during the 1990s as those structures came to dominate health care.) The future boundaries of bioethics are unpredictable; the changes will be to some extent organic, and to some extent a matter of historical accident. If the changes are suffciently radical, bioethics might not remain the appropriate label. To count as bioethics, I would argue, the activity must deal with moral questions that arise about health care, health policy, and the life sciences. Finally bioethics, like every other practice, is social. It cannot be learned well in isolation, because it is a matter of interacting with people and institutions as well as with ideas.

Defining History

MacIntyre pointed out that each practice has internal goals, understandable only in terms of the whole and partly constitutive of it. What are the bioethical equivalents of, for instance, checkmate?

This is a question about purposes and about distinctive ways of reaching them. The two are intertwined: the point of a race is not just getting to the finish line, but doing so in the way specified by the sport, whether that means swimming, skating, or running. One central goal of bioethics is to expand our understanding of the moral dimensions of health care and biological research, at the public as well as the individual level, and to help translate these ideals into practice. But we are out not only to understand but to make a difference; what most distinguishes bioethics from, say, academic philosophy is its relationship to the lived world. These purposes shape one another. Definitions of death, for instance, are not just clarifications but decisions, made not just in the light of physiological facts and conceptual coherence, but also in the light of what can be accepted by clinicians and patients and even what the implications for organ donation are.

The nature of our *intellectual* tools has been hotly disputed for years: should we be using principles, cases, narratives, all the above, or something else entirely? Our practical tools, with the single exception of ethics consultation within hospitals, have gotten little attention.

In the beginning, the academics who took up practical ethics thought of it as "applied ethics," solving ethical problems by applying ethical theory, especially general principles like those of utilitarianism ("Do whatever will produce the most happiness and the least suffering in the long run") or Kant's deontology ("Do only what you could consistently will that everyone do," "Treat human beings as ends in themselves, never merely as means"). The idea was to add to these principles whatever empirical information was available and then deduce an answer. For instance, keeping utilitarianism in mind, people asked what happiness and suffering would result if we allotted a scarce life-saving technology (like the original kidney dialysis machines) according to criteria of social worth, choosing, say, the mother of six over the town drunk. Questions like these are enlightening. (Allocating according to social worth, to continue the example, would foster a humiliating anxiety: "If I get sick, will be I be judged worthy of help?" This would be a significant kind of suffering.) Only a convinced utilitarian would take this line of inquiry as final, but almost everyone thinks that questions about consequences, about happiness and suffering, are important parts of ethical inquiry.

But even twenty-five years ago, the "apply the principles" model did not capture what practical ethics was like, not even if we confine ourselves to the activities of philosophers. For one thing, few people are completely convinced utilitarians or Kantians; even those who do not find their principles leading to simple, clear answers. In addition, the field has always used a variety of kinds of inquiry. Appeals to consistency, for instance, are commonplace in ethical analysis; moral agents require it of themselves. These appeals take the familiar forms of generalizability (What if everyone did that?), reversibility (Would you want that done to you?), and analogy (as in Judith Jarvis Thomson's famous abortion-and-the-violinist article). Moreover, basic concepts need, and from the beginning were getting, analysis: just what is privacy, for instance, and why does it matter?

Furthermore, the most important work of practical ethics done in the twentieth century, John Rawls's *A Theory of Justice,* explicitly endorsed what he called "reflective equilibrium": a moving back and forth between attractive general principles, on the one hand, and possibly conflicting convictions about particular cases on the other, checking and revising each level of conviction in the light of the other. *A Theory of Justice* was published in 1971, when bioethics was in its infancy.

So practical ethics was never simply an application of principles. Yet the label "applied ethics" stuck for a long time. Bioethics had its own specific version of this story: a misleading label took on a life of its own and remains today a target of attack (in my opinion, pointless attack). The story began in 1978 when Tom Beauchamp and James Childress published *Principles of Biomedical Ethics;* now in its fifth edition, it remains a leading textbook in the field. That landmark first edition aimed at bringing coherence to a field where discussion seemed scattered among many separate problems. As a result "the moral judgments involved in one dilemma appeared to be unconnected to the moral judgments in others." The authors argued that a small set of principles underlay all the particular dilemmas, both giving rise to them and being useful in resolving them. Beauchamp and Childress went on to list four principles eventually known (usually derisively) as "the Georgetown mantra." (Both men then taught at Georgetown University.) The principles were autonomy (the right of competent patients to make their own decisions about health care); non-maleficence (the duty of clinicians not to harm their patients); beneficence (the duty to help one's patients); and justice (a duty to distribute goods in short supply fairly). The authors begin authoritatively: "This book offers a systematic analysis of *the* moral principles that should apply to biomedicine". Their language invited the misinterpretation that persists today, that the authors were offering an exhaustive set of principles from which moral conclusions in bioethics follow deductively.

Yet, I would argue, this is not really what they were doing. The very paragraph that begins so authoritatively goes on to quote Tom Stoppard: "There would be no moral dilemmas if moral principles worked in straight lines and never crossed each other." Principles in a true deductive system *would* work in straight lines; they could not contradict one another. A few sentences later the authors write, "Only by examining moral principles *and how they conflict* can we bring *some* order and coherence to the discussion". "Some order and coherence" is a much more modest goal than is complete and deductive certainty. In fact, it would have been impossible for anyone trained in philosophy after about 1950 to aim at the latter. Our training centered in epistemology, in the nature and especially the limits of human knowledge, in the failure of various quests for certainty across the centuries and throughout intellectual life. We were also trained, however, to pursue understanding in spite of uncertainty, trying to identify our assumptions, work carefully, and keep our conclusions cautious. Of course we sometimes fail in this. But our professional commitment is to careful, skeptical inquiry, as opposed, say, to the exuberant nihilism of some forms of postmodernism.

Beauchamp and Childress, however, were not speaking primarily to philosophers. Their audience included clinicians, who by the nature of their jobs need clear guidelines, and tended to interpret "principles" as rules. Their audience also included academics from many different fields, who had no reason to recognize the systematic doubt assumed as background. In addition, there are a number of statements in the book that invite being read as flat assertions. The result was a misinterpretation ("principlism") that soon became, and remains today, a target of attack. Bioethics has fallen prey to a common academic fault, in which certain criticisms and constructions remain conventional wisdom long after they have ceased to apply, and even if they were never really accurate.

If practical ethics has never been simply the application of moral principle to concrete problems, then what is it? A more plausible position describes it as *moral reasoning* about practical questions: another kind of application, but considerably more open than the "application of principles" model. This interpretation, too, gives philosophy an important role. Philosophers can articulate the role that consistency plays in moral reflection, and are familiar with the sobering implications of some initially attractive principles. They should be skilled at clarifying concepts and setting out arguments. They bring the resources of our intellectual tradition to the table. (Thinking is often deeper when it takes into account what Aristotle, Kant, or Habermas have to say. Once we acknowledge that Shakespeare, Buddha, and Weber belong on this list, it becomes clear that all the humanities contribute to moral reasoning and reflection.)

The description of the application of moral reasoning to practical questions is better. But it is not enough. It describes classroom teaching and academic writing and ignores the rest of what makes up the field. Even as a description of intellectual activity it is misleading, since it suggests that fields like law, epidemiology, and sociology are subordinate: research assistants who dig out relevant information rather than colleagues in framing and in answering the questions. The description also pushes doctors and nurses into something like a traditional patient role, people to be improved, possibly in spite of themselves. In short, the "applied moral reasoning" approach does not capture the concrete, collaborative, and interdisciplinary nature of the field.

William Ruddick was one of the first to see things differently. In 1983, when "applied ethics" was still the dominant label, Ruddick identified "discursive moral competence" as a goal of ethics education: "the ability to discuss in appropriate moral terminology a variety of routine and rare cases with the variety of people likely to be involved in these cases physicians, nurses, patients, relatives, lawyers." His article fell by the wayside, but now seems prescient. Long before "discourse" came into intellectual fashion, Ruddick understood that one of our primary goals is to make it possible for people to talk with one another. Although this demands more than reasoning, and more than helping others improve their reasoning, it is not surprising that twenty years ago Ruddick focused on intellectual rather than social and psychological preconditions for productive conversation. Yet even there he was ahead of his time, understanding ethics talk more broadly than some still do. Ruddick identified "at least *three* modes of moral reasoning, " which he simplified as "the Protestant which stresses conflict of values and agonizing decisions; the Catholic mode, which supplies principles and the Jewish mode, which uses anecdotes (actual or fictional cases) to pose questions and suggest answers." (A twentyfirst-century version of this list would probably include Islam and Buddhism. Ruddick's point was not theological, of course, but practical. Moral discussion takes a variety of forms, however these are named.)

Two recent discussions are still richer. Neither is purely theoretical; each turns to metaphor to capture the work of practical ethics. The first comes from Margaret Urban Walker, the second from Caroline Whitbeck. Walker's topic was not the field as a whole but ethics consultation. She described it as "a kind of interaction that invites and enables something to happen, something that renders authority more self-conscious and responsibility clearer. It is also about . . . maintaining a certain kind of reflective space (literal and figurative) within an institution, within its culture and its daily life." Her point can be seen as a development of Ruddick's. He had allowed a variety of ways of processing ideas, of finding insight. For her, our goals include not just thinking ("ethics as moral philosophy") and talking ("ethics as discourse") but also awareness (of responsibility) and disposition (for reflection). And in linking metaphorical space to literal space, she moves bioethics outside the mind and into the world.

That last comment may seem unfair. Bioethics from the beginning has aimed at making a difference, in improving the way patients and research subjects are treated. But my second source, Caroline Whitbeck, points out that we often *describe* ethics as if it were a form of adjudication. In fact, she says, practical ethics is more like design. Judges rule on the rightness or wrongness of what someone else has devised, where designers try to create solutions. Each situation presents a distinct set of demands, and the job of the designer is to create a solution that best meets those with which she is faced.

Although there are clearly wrong answers, there is rarely a unique right answer. Anyone experienced in consultation, policy development, or exemplary practice knows that it is a form of problem solving, that ethical verdicts are never the whole point and sometimes simply beside the point.

Whitbeck's analogy could be further developed. For one thing, a designer must be responsive to parameters of many different kinds: the customer's desires, governmental regulations, the possibilities and limits of the physical world. An excellent designer will see possibilities and problems no one has thought to ask about. So in practical ethics we must be responsive to our client's desires, but also to the relevant law, and to morality. Our work goes beyond doing what is asked of us, a point to which I will return more than once. And designers do not just work on paper. They can create prototypes, look at problems in construction, and take a hand in designing the tools that will bring about the results they have imagined.

"Morality," like most terms, has a clear core and some disputed boundaries. Obviously it refers to the rightness and wrongness of actions, the virtues and vices of persons, the justice and legitimacy of political structures. I think of the moral attitude as fundamentally a matter of respect, but I do not use that term as Kant did. Respect, for persons and for the rest of the world, is a matter of recognizing, protecting, and nourishing whatever is of value. Under this description, morality cannot be adequately understood through philosophy alone. Equally essential are the other humanities and the sciences. Each in its own way helps us see what exists and understand its value. These, too, are points to which I will return in later chapters.

Practical ethics, then, is by nature interdisciplinary. The ability to see what is, to imagine what is not, to understand whether things can be changed no single discipline has a lock on so much. The ethical thing to do has been defined as that which should be done, all things considered. This is a large and powerful phrase. During that week in November 1994, I both drew from philosophy and grew as a philosopher, but I also called upon the little I knew of anthropology, with its constant warning that descriptions are suspect. (Anthropologists sometimes call themselves "natural anti-scripturalists.") When I began each consult, I knew that the initial description I was given would bear only a rough resemblance to the fuller story. I had to know something about biology and medicine, as well. Some knowledge of law (which I had) and of mediation techniques (which I did not) would also have been relevant.

Domain of Bioethics

Keeping moral space open, providing language and skills within it, identifying moral problems and helping create solutions for them: let us take these as a rough characterization of the goals that define bioethics.

For now, however, this loose preliminary description will suffice. The next task is to sketch the boundaries of our subject matter. After all, every field of practical ethics, whether in business, engineering, law, politics, or other fields, tries to maintain moral space, foster respectful conversation, and so on. But each has a distinct domain. For bioethics, this domain has largely been health care, health policy, and biological research.

The three arenas present an endless variety of moral questions from which bioethics has carved out a subset with an implicit central focus: the interactions between doctor and patient and researchers and their subjects. From the beginning I have been struck by the relatively narrow focus in bioethics, by the invisibility of whole categories of moral issues. Later I will call attention to the reasons for these boundaries and challenge their adequacy. At this point I want only to map them.

Bioethics tends to center around the doctor-patient interaction. I say "doctor" advisedly. Nurses do important work within bioethics, but much of their attention, too, focuses on what physicians should do. This discussion can empower nurses, who must occasionally, in conscience, resist doctors. But little in bioethics looks at nursing in its own right, as a separate field with its own problems. (In fact, most probably find that last sentence nonsensical. They cannot imagine nurses facing problems different in kind from those doctors face. "Competent patients have the right to refuse treatment; how could it matter which field is providing the treatment?")

Bioethics in the "dilemmatic" mode has focused on such questions as how much to tell patients, how strictly to keep their secrets, and especially, in every possible permutation, whether to allow or to hasten death. Literary scholars have examined the structure of medical records and case reports. Exemplary practices are usually explorations of ways to help dying patients. The methodological disputes of the past ten or fifteen years ("principlism" versus "narrative" versus casuistry, etc.) push the discussion one level back: rather than asking, "What should doctors do?, " it asks, "What tools are most helpful in figuring out what they should do?"

One kind of inquiry moves away from dilemmas and principles of action to deal with the subjectivity of the patient and of the doctor (again, I use the word advisedly; there is astonishingly little on the inner lives of nurses). What is it like to be sick, to suffer pain and disability, to face death? Conversely, what is it like to open bodies, catch babies, to see the disease in a patient who feels well? Explorations of subjectivity aim at helping doctors better understand themselves and their patients.

Societal issues addressed by bioethics usually involve access and allocation: should the government ensure universal health care, and if so how, and how should we distribute our finite resources? Often access to health care is treated as access to a doctor, but certainly not always; and what is scarce may not be doctors but technology or organs. Within the past few years, attention has gone to "managed care": here the frustrations of doctors get attention, as do policies that limit patients' access to doctors, technology, and medication.

The ethics of biological research began somewhat differently, centered in the rights of subjects to be protected *from* (rather than to be included in) research; recently the discussion has broadened to include the latter.

In all these areas, discussions take place at different levels. In each area there is discussion of particular cases: Should Carlos's sister be told he has AIDS? Were the placebo trials held in Africa justified? Should this child's liver transplant be funded? In each, there are also more general and theoretical discussions: What grounds the obligation of confidentiality, and how strong is it? What metaphors do people use as they describe being sick? What principles should we use in deciding who gets help? Discussion at any of these levels shapes discussion at others.

There are other subjects as well, flowing naturally from the field's historic core and changing its boundaries as they accumulate. New subjects arise from scientific advances (most recently in genetics), from the personal interests of influential people, and from repeated if dim glimpses of the fact that health care is not the only significant factor in health, nor doctors the only providers of care.

At the moment, then, the domain of bioethics centers upon health care, paradigmatically understood as a doctor helping a patient, but in principle including other health-care professionals as well. It includes the political and economic forces that shape medicine and access to it, and the biological research that underlies it. The domain shifts over time, growing in one area, shrinking in another, both acquiring and discarding foci. I will argue later that we need to take responsibility for this process, that we need to consider, and to reconfigure, the borders within which we operate.

Internal Goods, External Goods and Virtue

In addition to the complex of goals and means that constitute a practice, each also makes possible unique forms of excellence. Chess players can conduct brilliant end games; architects can design buildings beautifully suited to their function. The equivalents within bioethics include a well-done ethics consultation; a soundbite for journalists that makes exactly the right points and does it memorably; an effective course design; a task force that does its job well; an article that is morally sound and practically useful; a good palliative-care policy; and so on.

In addition, there are what MacIntyre calls "external goods." His own example is that of a child who has been offered 50 cents for each game of chess she plays. Learning to play for the sake of the reward, she eventually takes pleasure in the game itself. Money is an external good; that is, it can be gotten without playing chess, and is not a necessary part of the game. Many other kinds of reward also count as external goods: they are not part of the activity, although their pursuit can promote excellence in it, as in the case of the child learning chess. More to the point, external goods can endanger the specific excellences of an activity. This double role is often conspicuous in professional sports: On the one hand, money and fame help motivate athletes to high levels of achievement. On the other hand, money and fame can lead athletes to destroy the goods that are meant to be part of sport. Competitors will sometimes cripple one another or themselves in order to win, or will accept a payoff and "throw" a game. External goods can destroy internal ones.

Bioethics offers some significant external goods. When Ron Carson and Chester Burns asked whether twenty-five years of bioethics have made the sick better off, my instinctive response was, "I don't know. But *we* certainly are." The ways in which bioethicists benefit from doing bioethics deserve attention.

All professions offer some external goods, typically status, security, and a reasonable income. In most professions these rewards create at least some tension with internal goods. Junior faculty, for instance, can feel torn between the need to satisfy a promotion committee (an external good) and the internal goods of excellent teaching and research. For there can be a difference, unfortunately, between trying to please students and trying to teach them, even though the two may be closely linked. Along the same lines, publishing for the sake of promotion is different from pursuing knowledge for the sake of the common good. Physicians also face conflicts: in a fee-for-service system, some will order tests or procedures not because a patient needs them but because they get paid more every time they order one. The motivation may be conscious or unconscious.

Bioethics offers higher salaries and larger stipends than do the standard academic liberal arts. As a newcomer to the field, I was struck by its relative affluence. During my first year in the field, I was invited to speak at another college; the arrangers asked what my stipend would be. I didn't know what to say, so I asked a colleague: "Would $35 be about right?" He was astonished. "Don't do that! You'll depress the market!" So, upon his advice, I asked for, and got, $350. For the "service" components of our work we often receive stipends five or ten times what a philosopher of comparable stature might ask. This comparison does not hold for medicine. If doctors are right in thinking that bioethics lowers their salaries, these particular external goods cannot threaten their integrity. The danger does exist for humanists.

In addition, bioethics can offer a certain degree of celebrity, if one is quoted in the media or does much public speaking. We can also have a certain amount of power, about specific cases or about policy, although the power is always diluted: our role is advisory, and usually we serve as members of a committee or a board. Finally, of course, there is the allure of security (in the form of tenure, when it is available) and of status within the profession.

So conflicts of interest of varying degrees are built into bioethics, as they are in all professions. The chance to gain money, security, prominence, or influence could distort what one chooses to say or do. This is obvious in dealing with the press, subtler in other situations. My colleagues and I have written elsewhere about interactions with the media. Later parts of this book will explore some of the other conflicts. My major reason for identifying these external goods, however, is not to offer solutions for particular conflicts but to help in identifying virtues that may be of particular importance in bioethics, in part because of competing external goods.

MacIntyre's framework raises the possibility that individual practices demand distinctive virtues. (Something like "resisting countertransference, " for instance, is probably a virtue necessary for psychotherapists, and not even nameable outside that practice.)

In exploring the role of virtue in the practice of bioethics, I will make no attempt to give a list. But I will argue that the lure of external goods highlights the importance of integrity and discernment, and that interdisciplinary work particularly requires intellectual virtues. I will describe an overlooked virtue that I can label no more simply than "choosing projects well." Finally, I will discuss how the communities we form contribute to or interfere with these virtues.

Purposive Practices

I asked almost all of those whom I interviewed whether they thought bioethics made any real difference in the world. "Yes!" said one, vehemently. Then, after a brief pause, she said more quietly, "It had better make a difference." There was another, longer pause. Finally, more cautiously, she ventured, "A lot of people don't think about things till it's too late. Patients, family, and staff." The implication was that her work helped them take up important questions earlier, when it might help. To my question another person said simply, "I have to believe that it does" make a difference. Almost all the responses I got fit within this spectrum. Often the answer was a kind of assertion of faith, and certainly a recognition that, unlike some of the arts, other humanities, and even sciences, bioethics is a failure unless to some extent it improves the world.

MacIntyre, writing about practices in general (certainly including the arts), takes a more modest position. He suggests that practices should fit together, each contributing in some way to the possibility of full, good human lives, if only for those who engage in them. David Miller, however, in response, makes a distinction: some practices have as an *internal good* the production of extrinsic goods: agriculture must produce food, medicine must, in some sense, heal. This response provides another tool for the moral evaluation of bioethics: does it accomplish anything, and is what it accomplishes good?

These are large questions. Some of them I will address in this book, but others demand broad-scale empirical investigation and are beyond my reach. Some of the initial findings are not encouraging. A number of large recent studies, for instance, suggest that "advance directives" ("living wills" and assignments of powers of attorney for health care) make little difference in health care. These documents have been advocated for years as ways of changing some undesirable facts about dying in America: there is too much pain, certainly, and probably more intervention than patients want. On the other hand, the findings themselves, together with the debate about physician-assisted suicide, have at last led to serious attention to the way health care handles death, and the landscape may finally change for the better not for the wrong reasons, exactly, but because of an unexpected sequence of causes. Questions about whether we make a difference are complex.

Although almost everyone to whom I talked believed bioethics makes some difference, most also recognized that we don't really know. The one clear exception was someone who did quantitative work. He answered without hesitation. "Oh, yeah, bioethics has made a difference. The FDA's `Fast Track' decisions, for instance: bioethics gave a voice to the AIDS activists; they needed legitimacy from people outside the movement. And bioethics kept informed consent alive even when the courts were backtracking. Advanced directives, too. We have data showing how much the families felt relieved, knowing they had the advance directive to support their decisions."

Many who were less concrete, who said, in essence, "I have to think we make a difference," had teaching in mind. For many of us, teaching is fundamentally an act of hope. Year after year students pass through our classrooms and on into their lives. Sometimes, years later, we learn that those classroom hours mattered. Usually we never know.

Another sort of response was also common. "We contribute to a lively public debate. Sometimes we even create that debate. And that, in turn, shapes general awareness, policy, customs and individual choices." I believe this thinking may well be right, and that we are mistaken to evaluate the impact of bioethics solely by looking at some of its specific activities. The question is not just whether, say, ethics courses change the medical students who take them, but whether the now universal presence of ethics courses changes the fabric of assumptions in which medicine is practiced. Similarly, an ethics consultation service might not make a difference directly; the results in cases where consultation is sought might not differ statistically from the usual hospital pattern. But it is possible that patterns of practice in the hospital will change in the years the service exists, because of its sheer presence, the individual consultations it provides, and its educational activities. Obviously specific activities need to be assessed, and here again the preliminary findings are disquieting. Ethics consultations may reinforce medical power rather than challenge it, for instance.

The same kind of question arises about our overall impact. If we are, for instance, creating and sustaining public debate, is the shape of that debate good? My doubts about the answer began during that week in November 1994, when I first suspected that bioethics helps divert attention from poverty, the most serious health problem in the United States. Europeans sometimes comment upon the American fascination with health care, in contrast to the considerably cooler attitude in Europe. I have come to believe that bioethics both feeds upon and feeds that fascination. Henk ten Have believes Americans' interest will cool. Until that happens, and I see no signs that it will, several things concern me. One is the valorizing of health and "health behaviors," which sometimes seem our only secular values, and ones that are used to validate everything else. (I've seen friendship, sensuality, spirituality, and altruism all praised because they seem to contribute to good health.)

Bibliography

Albrand, rapporteur, *Rapport de la Commission Spéciale des Docks au Conseil Municipal de la Ville de Marseille* (Marseille: Typographie des Hoirs Feissat Ainé et Demonchy, 1836).

Albury, Randall, *The Politics of Objectivity* (Victoria, N.S.W.: Deakin University Press, 1983).

Alcouffe, Alain, "The Institutionalization of Political Economy in French Universities, 1819–1896," *History of Political Economy,* 21 (1989), 313–344.

Alder, Ken, "A Revolution to Measure: The Political Economy of the Metric System in France," in Wise, *Values of Precision,* 39–71.

Barber, William J., *From New Era to New Deal: Herbert Hoover, the Economists, and American Economic Policy, 1921–1933* (Cambridge, U.K.: Cambridge University Press, 1985).

Barnes, Barry, "On Authority and Its Relation to Power," in Law, *Power,* 180- 195.

Bartrip, P.W.J., and P. T. Fenn, "The Measurement of Safety: Factory Accident Statistics in Victorian and Edwardian Britain," *Historical Research,* 63 (1990), 58–72.

Baum, Charles, "Des prix de revient des transport par chemin de fer," *Annales des Ponts et Chaussées* [5] 10 (1875), 422–481.

Carson, John, "Army Alpha, Army Brass, and the Search for Army Intelligence," *Isis,* 84 (1993), 278–309.

Carter, Luther J., "Water Projects: How to Erase the 'Pork Barrel' Image," *Science,* 182, October 19, 1973, 267–269, 316.

Cartwright, Nancy, *Nature's Capacities and Their Measurement* (Oxford: Clarendon Press, 1989).

Caufield, Catherine, "The Pacific Forest," *New Yorker*, May 14, 1990, 46–84.

Daston, Lorraine, and Katherine Park, *Wonders of Nature: The Culture of the Marvelous, 1500–1740* (Cambridge, Mass.: Harvard University Press, forthcoming).

Davis, Audrey B., "Life Insurance and the Physical Examination: A Chapter in the Rise of American Medical Technology," *Bulletin of the History of Medicine*, 55 (1981), 392–406.

Day, Archibald, "On the Determination of the Rates of Premiums for Assuring against Issue," *Assurance Magazine*, 8 (1858–60), 127–138.

Day, Charles R., *Education for the Industrial World: The Ecoles d'Arts et Métiers and the Rise of French Industrial Engineering* (Cambridge, Mass.: MIT Press, 1987).

Etner, François, *Histoire du calcul économique en France* (Paris: Economica, 1987).

Evans, Hughes, "Losing Touch: The Controversy over the Introduction of Blood Pressure Instruments into Medicine," *Technology and Culture*, 34 (1993), 784–807.

Farren, Edwin James, "On the Reliability of Data, when tested by the conclusions to which they lead," *Assurance Magazine*, 3 (1852–53), 204–209.

Faure, Fernand, "Observations sur l'organisation de l'enseignement de la statistique," *Journal de la Société de Statistique de Paris*, 34 (1893), 25–29.

Galambos, Louis, ed., *The New American State: Bureaucracies and Policies since World War II* (Baltimore: Johns Hopkins University Press, 1987).

Galison, Peter, "Aufbau/Bauhaus: Logical Positivism and Architectural Modernism," *Critical Inquiry*, 16 (1990), 709–752.

Geison, Gerald L., ed., *Professions and the French State, 1700–1900* (Philadelphia: University of Pennsylvania Press, 1984).

Habermas, Jürgen, *The Structural Transformation of the Public Sphere*, Thomas Burger, trans. (1962; Cambridge, Mass.: MIT Press, 1989).

Hackett, John, and Anne-Marie Hackett, *Economic Planning in France* (Cambridge, Mass.: Harvard University Press, 1963).

Jasanoff, Sheila, "The Problem of Rationality in American Health and Safety Regulation," in Smith and Wynne, *Expert Evidence*, 151–183.

Jasanoff, Sheila, *Risk Management and Political Culture* (New York: Russell Sage Foundation, 1986).

Jasanoff, Sheila, *The Fifth Branch: Science Advisers as Policymakers* (Cambridge, Mass.: Harvard University Press, 1990).

Kang, Zheng, "Lieu de savoir social: La Société de Statistique de Paris au XIXe siècle (1860–1910)," Thèse de Doctorat en Histoire, Ecole des Hautes Etudes en Sciences Sociales, 1989.

Keller, Evelyn Fox, *Reflections on Gender and Science* (New Haven, Conn.: Yale University Press, 1985).

Labiche, Eug., and Ed. Martin, *Les Vivacités du Capitaine Tic* (Paris: CalmannLevy, n.d.; first performed 1861).

Labry, Félix de, "A Quelles conditions les travaux publics sont-ils rémunerateurs," *Journal des économistes*, 10 (November 1875), 301–307.

Mauskopf, Seymour, and Michael R. McVaugh, *The Elusive Science: Origins of Experimental Psychical Research* (Baltimore: Johns Hopkins University Press, 1980).

Meehl, Paul E., *Clinical versus Statistical Prediction* (Minneapolis: University of Minnesota Press, 1954).

Megill, Allan, "Introduction: Four Senses of Objectivity," in Megill, *Rethinking Objectivity*, 301–320.

Nathan, Richard P., *Social Science in Government: Uses and Misuses* (New York: Basic Books, 1988).

Navier, C.L.M.H., "De l'exécution des travaux, et particulièrement des concessions," *Annales des Ponts et Chaussées*, 3 (1832), 1–31.

O'Malley, Michael, *Keeping Watch: A History of American Time* (New York: Viking, 1990).

Orlans, Harold, "Academic Social Scientists and the Presidency: From Wilson to Nixon," *Minerva*, 24 (1986), 172–204.

Patriarca, Silvana, *Numbers and the Nation: The Statistical Representation of Italy, 1820–1871* (Cambridge, U.K.: Cambridge University Press, forthcoming).

Pauly, Philip, *Controlling Life: Jacques Loeb and the Engineering Ideal in Biology* (New York: Oxford University Press, 1987).

Ross, Dorothy, *The Origins of American Social Science* (Cambridge, U.K.: Cambridge University Press, 1991).

Schaffer, Simon, "Late Victorian Metrology and Its Instrumentation: A Manufactory of Ohms," in Robert Bud and Susan E. Cozzens, eds., *Invisible Connections: Instruments, Institutions, and Science* (Bellingham, Wash.: SPIE Optical Engineering Press, 1992), 23–56.

Schiesl, Martin J., *The Politics of Efficiency: Municipal Administration and Reform in America* (Berkeley: University of California Press, 1977).

Thévenez, René, *Legislation des chemins de fer et des tramways* (Paris: H. Dunod et E. Pinat, 1909).

Thévenot, Laurent, "La Politique des statistiques: Les Origines des enquêtes de mobilité sociale," *Annales: Economies, sociétés, civilisations*, no. 6 (1990), 1275–1300.

Von Mayrhauser, Richard T., "The Manager, the Medic, and the Mediator: The Clash of Professional Styles and the Wartime Origins of Group Mental Testing," in Sokal, *Psychological Testing*, 128–157.

Wagner, John W., "Defining Objectivity in Accounting," *Accounting Review*, 40 (1965), 599–605.

Index